Managing Changing Prairie Landscapes

Managing Changing Prairie Landscapes

edited by

Todd A. Radenbaugh and Glenn C. Sutter

2005

Canadian Plains Research Center
University of Regina
Regina, Saskatchewan S4S 0A2
Canada
Tel: (306) 585-4758
Fax: (306) 585-4699
e-mail: canadian.plains@uregina.ca
http://www.cprc.uregina.ca

Library and Archives Canada Cataloguing in Publication
Managing changing prairie landscapes / editors: Todd A. Radenbaugh, Glenn C. Sutter.

(Canadian plains proceedings ; 35)
Articles originating in the Plain as the Eye Can See Public Forum/Conference, held in Regina, May 15-17, 2003.
Includes bibliographical references and index.
ISBN 0-88977-177-4

1. Landscape–Prairie Provinces–Congresses. 2. Land use–Prairie Provinces–Planning–Congresses. 3. Ecosystem management–Prairie Provinces–Congresses. 4. Ecology–Prairie Provinces–Congresses. 5. Nature–Effect of human beings on–Prairie Provinces–Congresses. I. Radenbaugh, Todd A., 1964- II. Sutter, Glenn C. (Glenn Christian), 1963- III. University of Regina. Canadian Plains Research Center IV. Plain as the Eye Can See Public Forum/Conference (2003 : Regina, Sask.) V. Series.
GF512.P7M35 2005   333.73'09712
C2004-906797-4

Printed and bound in Canada by: Houghton Boston, Saskatoon
Cover design by Donna Achtzehner, Canadian Plains Research Center
Cover photograph courtesy David McLennan

Index prepared by Adrian Mather (amindexing@shaw.ca)
We acknowledge the financial support of the Government of Canada through the Book Publishing Industry Development Program (BPIDP) for our publishing activities.

CONTENTS

CHAPTER 1

The Challenges of Managing Changing Prairie Landscapes

Todd A. Radenbaugh and Glenn C. Sutter

Introduction

This volume is the result of a free public conference that took place at the Royal Saskatchewan Museum on May 15–17, 2003. The conference—the third of a series called "Plain as the Eye Can See"—focused on management policies that affect northern prairie ecosystems (Plate 1). The goals of the conference were to explore the effects of current management strategies on resources in the northern Great Plains and to examine possible futures using different types of capital and full cost accounting as central themes. To stimulate thinking and focus discussion, participants were asked to consider the following questions:

• Does living in a dynamic landscape have any policy implications?

• How should/could changing landscapes be managed so they remain sustainable or healthy?

• How can a balance be achieved between social, economic and ecological concerns?

• Where should limited dollars be spent to reach a defined management goal?

• What barriers or shortfalls are hampering the development of healthy communities?

The conference began with a keynote address entitled "Did the Hills Move? Science and Spirituality" by noted prairie author Sharon Butala. Through stories based on personal experience, she spoke about the intrinsic and aesthetic value of undisturbed prairie landscapes. She reminded us that there is much about these landscapes that is forever unknowable and beyond our control—two features that are usually overlooked in positivist, techno-centric management plans.

By calling for humility in our views and use of the land, Butala was echoing a well-established fact about prairie landscapes and other ecosystems—they are always changing, often in unpredictable ways. Today's northern prairie landscape began to form over 15,000 years ago as the most recent Pleistocene glacier melted, and it has been changing ever since. The short-term climatic cycles that affect the area, such as drought, flood, unseasonable cold periods, and heat waves, all happen within a broader pattern of regional warming that has been occurring for thousands of years. As these changes occurred, the biota and indigenous societies have changed in response, as they tracked their preferred climatic bands (Axelrod, 1985). Even structures that appear to be static in human terms, such as the region's geomorphology and soils, are constantly being changed through geological processes such as weathering and erosion.

As dramatic as these changes have been they have never been a cause for alarm, since the pace of change was never rapid enough to be witnessed in a human life-span or to cause a sustained drawdown on critical supplies of natural, social, human, and manufactured capital. What we should be alarmed about is the accel-erating pace of landscape-level change that has occurred since the arrival of rail-ways in the 1880s. In addition, there have been new causes of change, notably industrialized agriculture and urbanization, which are largely responsible for a decline in the native biota, and various social issues. These new landscape-level changes have altered the ecosystem to a point where it is struggling to provide the services people require (see list of ecosystem services in Appendix 1), and the future repercussions on society will be significant.

In terms of economic growth within the landscape, we have come to a point where we must recognize and respect limits if we want to sustain the ecosystem serv-ices the northern prairies provide. This is witnessed in our ecological footprint. Ecological footprints put the magnitude of our impacts in perspective by measur-ing the land and water required to provide us with goods and services, including food production and waste management (Wackernagel and Rees, 1995). A recent assessment (Venetoulis et al., 2004) indicates that our global footprint has been growing since the 1960s, and that our collective impact has been greater than the productive surface area of the Earth since the mid-1970s. In other words, our global society has been in a deficit situation for the last three decades, drawing down on critical sources of capital in order to sustain itself.

The situation may seem less dire on the northern prairies because human pop-ulation densities are low, but we should remember that the average North American has a relatively large footprint. According to Statistics Canada census data, there were about 5 million people in the Canadian portion of the Prairie Ecozone in 2003 (www.statcan.ca). Given that the average Canadian has an eco-logical footprint of 8.56 ha (Venetoulis et al., 2004), then the collective impact of this population amounts to almost 43 million ha, or over 95% of the available land area (45 million ha). The correlation between a population's footprint and its land base is not this direct, however, since the average Canadian uses resources from all over the world and the prairie farming and ranching economies are almost entire-ly based on export activities. Given that Canadian prairie economies are providing goods and services to other industrialized economies which are also producing large per capita footprints (Venetoulis et al., 2004), it is reasonable to assume that current prairie landscapes are stretched beyond their sustainable limits, despite low numbers of people.

Our ecological footprints are having detrimental influences on Earth's biologi-cal diversity and the functioning of biophysical systems (Costanza et al., 1997; Matson et al., 1997; Vitousek et al., 1997). Society is also affecting the composition and structure of species assemblages, changing the fundamental elements that define ecosystems (Abrams, 1996; Grime, 1999). These ecological influences and other social consequences raise questions concerning management actions that we ought to be taking to counter these pressures. On the northern prairies, we are faced with many sustainable management challenges that cross social, economic, and ecological disciplines, so we must consider not only the economic but also the social and environmental dimensions of our policies and actions.

The purpose of this issue is to explore ways we manage human activities to maintain a healthy northern prairie system. We recognize that there is no way to manage an entire landscape such as the northern prairie. The real management lies in complex biological and geological interactions that operate at scales of time and space (both micro and macro) that are beyond the human experience. What humans *can* do is add or remove components, alter fluxes and relationships, and change various resource reservoirs. If sustainable use of a landscape is desired, then the most efficient way to achieve this goal is to manage how society interacts with a landscape. To do so calls for adaptive management based on modesty, and ecocentric strategies that recognize our dependence on local, regional, and global ecosystems (Mosquin and Rowe, 2004).

Sustainability

Sustainable development, or sustainability, is a challenging, multi-dimensional concept with almost 60 working definitions (Murdock, 1997). The central goal is to ensure that economic activities do not prevent future generations from living in healthy ecosystems. At its core, sustainability involves respecting ecological constraints, giving present and future generations a fair opportunity to live fulfilling lives, working for social justice, developing a steady-state economy, and appreciating the complexity of social and ecological systems (Sutter and Worts, in press). This broad, interdisciplinary view involves integrating ecological, geological, societal, and economic dimensions that operate within multiple spatio-temporal scales, a challenge promoted in the 1980s by the Brundtland commission (World Commission on Environment and Development, 1987). This approach is echoed by Marten (2001), who argues that sustainable systems require the blending of ecosystem health, economic development, and social justice as these elements are all mutually reinforcing.

To live sustainably, society needs to manage its activities in ways that consider the ebb and flow of capital, including manufactured capital (consumer goods, buildings, transportation, and communication networks); natural capital (organisms, space, raw materials, and clean air, water, and soil); human capital (knowledge and technology); and social capital (human interactions and social cohesion). From a systems perspective, natural capital passes through a four-phase cycle defined by degrees of storage and connectivity (Holling, 1992 and Fig. 1). The cycle's flux is most rapid in the release phase (e.g., a forest fire), where stored capital becomes available to the system as a whole. This is followed by reorganization (invasion by pioneer species) where capital is placed back into biotic reservoirs but the connectivity of the system is relatively low. Next comes a slow exploitation phase (succession) where more complex ecological relationships develop and the amount of stored capital increases in the biosphere. Finally, the system enters a conservation phase (climax) with the highest biological reservoir of stored capital that is maintained by complex ecological interconnectedness. This phase is maintained until a disturbance initiates a new release phase (the magnitude of which may vary).

There are several important connections between Holling's (1992) model and sustainability work. First, other types of capital appear to exhibit a similar cyclical pattern. Some transitions in manufactured capital are easy to see, from the rapid decay of paper products to the growth of urban centres. The ebb and flow of social and economic systems is evident as institutions adapt or become extinct in the face

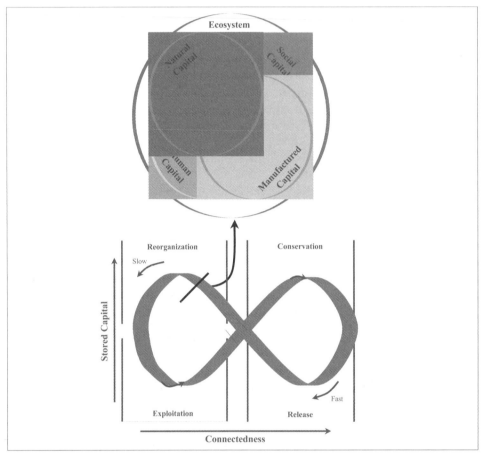

Figure 1. A conceptual model showing the phases and integration of capital flows in a human-dominated ecosystem. The stages defined by the storage and connectedness of capital and the figure-8 pathway are from Holling's (1992) model of ecological succession. The overlapping circles show the pathway in cross-section.

of consumer pressure and entire economies go through cycles of wealth concentration and redistribution. There are also transitions that affect human capital, loosely defined as the collective knowledge of a society. Theories of learning based on constructivism (Hein, 1998) suggest that people acquire understanding as individuals and groups by periodically deconstructing and rebuilding knowledge. Second, we can assume that capital types influence each other in a non-linear fashion, with fluxes ultimately constrained by the level of the systems through which they move, which can range from individual households to the global ecosphere. Finally, our modification of Holling's model suggests that, rather than trying to maintain systems in a certain state indefinitely, sustainability has more to do with ensuring that the fluxes operate and state transitions occur without reducing the welfare of future generations. In other words, sustainability is more complicated than conservation or preservation, and must have an appreciation for ecological constraints.

In this volume, we have tried to combine expertise from many disciplines, recognize and accommodate multiple viewpoints, and concentrate on more than one capital type, all the while realizing that landscape-level issues are multifaceted and hierarchical. While the result offers a mostly empirical approach appropriate for wildlife management and conservation policy, we recognize that there are alternative "ways of knowing" that need to be considered, from qualitative studies to traditional ecological knowledge.

Human Influences on the Northern Prairie Landscape

The magnitude of human alterations on northern prairie ecosystems raises concern about the how ecosystems change in terms of biophysical processes and system integrity. Studying the pressures and influences that society has on ecosystems can help in understanding the complex processes that operate within present landscapes. Furthermore, because of the substantial changes in the larger landscape, it is reasonable to question whether the ecological roles of biota, and the ecological services they provide, are quantitatively and qualitatively different today than they were prior to European settlement. The broad questions raised by this volume include:

- Have society's activities altered species resource use and, if so, are there changes in local or regional assemblages?
- What part of the human systems should we change to make society more sustainable?
- How much of an intrusion or perturbation has agricultural settlement been on native, naturalized, and introduced species and their use of resources?
- How has the conversion of significant land area into cropland influenced regional species diversity?
- What are the societal influences on ecosystem level processes?
- How might society adapt to the changing conditions in the landscape?

As with other regions, the northern Great Plains of North America have experienced many of the broad ecological influences of human habitation. Agricultural activities now play dominant roles within many ecosystems, controlling biophysical processes that affect ecosystem functions and even human health (Kaiser and Gallagher, 1997; Tilman, 1996; Tilman et al., 1997; Vitousek et al., 1997). Since the late 1800s, the northern prairies have lost approximately 80% of their native habitats due to their conversion into other land cover types and the remainder has been highly fragmented (Samson and Knopf, 1996; Selby and Santry, 1996). Agrochemical use and its addition of compounds to the soil, surface water, groundwater, and atmosphere may also have influenced the ecological processes of this region (Botsford et al., 1997; McMichael, 1997). While these landscape-level changes have multifaceted causes, cultivation and urban development are typically cited as primary forces (Saunders et al., 1991; Herkert, 1994; Joern and Keeler, 1995; Samson and Knopf, 1996; Radenbaugh and Douaud, 2000). In addition, two of the dominant ecological structuring forces in this region, namely fire (Rowe, 1969; Wright and Bailey, 1980; Collins and Wallace, 1990; Collinsm 1992) and large grazing ungulates (Arthur, 1984; Mitchell, 1984; Frank et al., 1998; Knapp et al., 1999), have been largely eliminated or are heavily managed in relation to agricultural activities. In less than two centuries, agricultural land-use in the northern prairies has left a matrix of ecological landscapes that have been structurally altered to various degrees.

Within the northern prairie landscape, conservation practices have been implemented in many areas that take into account the historically recognized natural processes of this landscape (e.g. fire and grazing) (Anderson et al., 1996; Knapp et al., 1999). These management measures, however, are generally conducted on only a small fraction of the total area and may not significantly counter the effects of agricultural development in the region (Peltzer, 2000). For example, actions to protect summer breeding grounds may not be enough to protect the many bird populations that migrate to other ecosystems in winter (Sherry and Holmes, 1996).

In This Issue

This collection of articles brings together a breadth of research that concerns natural, social, manufactured, and human capital. Although an individual article may not delve into all four capital types, each one tries to bridge at least two.

The first article by Riemer briefly outlines the settlement of the Saskatchewan prairies, then looks at the influence of recent land-use policy on waterfowl species. He concludes that the reduction in farmland in the 1990s is not enough to account for the rise in some bird populations and that land that has been reseeded to grass does not have the same ecosystem integrity as native prairie. He also notes that for most native waterfowl, wetlands are the key, and their conservation has become a "just add water" issue.

As the northern prairies were settled, society became increasingly dependent on non-renewable energy inputs. This dependency has increased labor efficiency, but at the cost of a significantly reduced labor force and the reliance on expensive equipment. Stirling examines the details of prairie agriculture's dependence on non-renewable energy in Saskatchewan and finds that the switch from animal to petroleum power did not significantly increase energy efficiency. He urges caution when promoting policies that further increase farm inputs.

The rural landscapes of Saskatchewan are further explored in Diaz and Nelson's paper on social cohesion, where they analyze the dynamics of social capital. Based on the prevalence of trust and high levels of participation in formal and nonformal organizations, they conclude that rural communities have a solid foundation of social capital that can help them adapt to recent environmental and socio-economic challenges.

Belcher and Schmutz concentrate on integrating capital types by proposing the Wood River Cooperative, where rural agricultural products are sold to urban consumers. By linking food producers directly with urban consumers, much of the waste in storage, marketing, and transportation is lost and consumers share in the cost of protecting ecosystem functions. This benefits the producer, consumer, and environment. Furthermore, urban consumers develop a direct connection to the land that produces the majority of their food, further promoting a sustainability ethic.

Sutter et al. examine the link between natural and social capital in community-based ecosystem management (CBEM) projects, using the Frenchman River Biodiversity Project as a case study. As natural capital, biodiversity provides a useful focus for CBEM projects because it is sensitive to land-management decisions and is easily monitored through community-based activities, but it can also be problematic. To foster mutual trust, project planners need to consider a range of issues, including landowner concerns about species at risk and the ability of researchers to publish results.

Jefferson et al. then explore links between natural and social capital through the use of native grasses to improve rangeland. When native, drought-resistant forage is coupled to standard grazing systems, native grasses offer huge potential in rangelands, increasing both natural and produced capital over decades. These native pastures also promote prairie wildlife and improve soil organic matter when compared to tame pastures.

More on the natural capital side, Hamilton proposes a novel method of both identifying remnant prairie patches and mapping the distribution of ecological regions. Using the distribution of specialized insect species with specific habitat requirements and small home rages, plots of remnant prairies can be classified and linked into ecoregions. This paper challenges some conventional wisdom concerning ecological classification and offers a useful tool for plotting pre-settlement vegetation.

Wolfe and Thorpe then investigate drought and landscape sensitivity on northern prairie landscapes, with an eye towards sandhills and the potential effects of climate change. As marginal landscapes that are sensitive to slight changes in climate, sandhills are good indicators of change. Future climate scenarios indicate that some sand fields may become active again, greatly affecting both the natural and human communities in the region. Furthermore, models of climate change suggest that plant growing conditions will change, altering species composition in these habitats.

Sauchyn et al. expand on this concept by looking at the potential for climate change over the entire Canadian Prairies and assessing if the region is prone to desertification. Their paper illustrates that producers and institutions have had, and may continue to have, the capacity to prevent desertification. However, with increasing levels of greenhouse gases, the risk of desertification persists and may increase. The authors also note that the landscape is sensitive to these changes and that society must adapt, as it would be difficult to alter current trends.

Radenbaugh discusses the various recommendations made by the authors with an eye towards managing the region as an ecosystem. The paper emphasizes that the regional biota in the northern prairie landscape have changed and will continue to do so, using historic changes in grassland bird populations as an example. Where grassland birds are concerned, the most notable changes include significant decreases in local native species diversity, but increases in regional species diversity. Thus the regional resource use of these animal populations can provide an indication of the extent of society's interaction with the landscape.

Conclusions

The behaviour of complex systems is often difficult to appreciate and understand. As mentioned above, each type of capital is part of dynamic, unrelenting cycles (Figure 1), and sustainability is about allowing these cycles to take place in ways that do not limit the ecological, social, or economic welfare of future generations. Through proper management, we can create opportunities to allow systems to cycle through peaceful and non-catastrophic means, and we need to proceed with humility. The tenets of systems ecology suggest that if something as complex as the climate system is disturbed past a threshold the system will start to move towards another stable state (see Sauchyn et al. this issue), no matter how society tries to prevent it. Furthermore, when ecosystem management is the goal, there are many types of "currency" besides money that need to be considered, such as wet-lands, clean air, carbon, energy, communication and road networks, social cohesion, and technology. We need to acknowledge these currencies and make better use of them through developing an environmental trading market such as outlined by Belcher and Schmutz (this issue) but on a global scale.

Ultimately, the changing prairie landscape will affect the health of its residents. Modelers suggest that the prairies will witness warmer temperatures, increased drought, and increases in air pollution over the next few decades (Sauchyn et al. this issue). These climatic changes will cause biotic changes as species track their preferred habitats, and the movement of species could in turn influence insect- or water-borne diseases that affect humans, their crops, and their livestock. Such changes will place more stress on the health infrastructure and social-support systems. To ensure sound management and a secure future, we need to recognize the range of possible futures humanity might face and develop plans that will start to foster suitable adaptations now.

Acknowledgements

We wish to thank all of the authors for their diligent work and patience as this issue was compiled. Transdisciplinary research and writing is often more time consuming and difficult due to the wider audience. We also thank the Prairie Adaptation Research Collaborative and the Canadian Plains Research Center at the University of Regina and the Royal Saskatchewan Museum for funding of the "Plain as the Eye Can See" forum. Special thanks go to Brian Mlazgar for all of his efforts in bringing the publication to print.

Literature Cited

Abrams, P.A. 1996. "Evolution and the Consequences of Species Introductions and Deletions." *Ecology* 77: 1321–1328.

Anderson, M.G., R.B. Fowler and J.W. Nelson. 1996. "Northern Grassland Conservation and the Prairie Joint Ventures." In Samson, F.B. and F.L. Knopf (eds.), *Prairie Conservation: Preserving North America's Most Endangered Ecosystem*. Washington, DC: Island Press.

Arthur, G. 1984. "The North American Plains Bison: A Brief History." *Prairie Forum* 9: 281–289.

Axelrod, D.I. 1985. "Rise of the Grassland Biome, Central North America." *The Botanical Review* 51: 163–201.

Botsford, L.W., J.C. Castilla and C.H. Peterson. 1997. "The Management of Fisheries and Marine Ecosystems." *Science* 277: 509–515.

Collins, S.L. 1992. "Fire Frequency and Community Heterogeneity in Tallgrass Prairie Vegetation." *Ecology* 73: 2001–2006.

Collins, S.L. and L.L. Wallace. 1990. *Fire in North American Tallgrass Prairies*. Norman: University of Oklahoma Press.

Costanza, R., R. d'Arge, R. de Groot R, S. Farber, M. Grasso, B. Hannon, K. Limburg, S. Naeem, R.V. O'Neill and J. Paurelo. 1997. "The Value of the World's Ecosystem Services and Natural Capital." *Nature* 387: 253-260.

Frank, D.A., S.J. McNaughton and B.F. Tracy. 1998. "The Ecology of the Earth's Grazing Ecosystems." *Bioscience* 48: 513–521.

Grime, J.P. 1997. "Biodiversity and Ecosystem Function: The Debate Deepens." *Science* 277: 1260-1261

Hein, G.E. 1998. *Learning in the Museum*. New York: Routledge.

Herkert, J.R. 1994. "The Effects of Habitat Fragmentation on Midwestern Grassland Bird Communities." *Ecological Applications* 4: 461–471.

Holling, C.S. 1992. "Cross-scale Morphology, Geometry, and Dynamics of Ecosystems." *Ecological Monographs* 62, 447–502.

Joern, A. and K.H. Keeler. 1995. *The Changing Prairie: North American Grasslands*. New York: Oxford University Press.

Kaiser, J. and R. Gallagher. 1997. "How Humans and Nature Influence Ecosystems." *Science* 277: 1204–1205.

Knapp, A.K., J.M. Blair, J.M. Briggs, S.L. Collins, D.C. Hartnett, L.C. Johnson and E.G. Towne. 1999. "The Keystone Role of Bison in North American Tallgrass Prairie." *Bioscience* 48: 39–50.

Marten, G.G. 2001. *Human Ecology: Basic Concepts for Sustainable Development*. Sterling, VA: Earthscan Publications Ltd.

Matson, P.A., W.J. Parton, A.G. Power and M.J. Swift. 1997. "Agricultural Intensification and Ecosystem Processes." *Science* 277: 504–509.

McMichael, A.J. 1997. "Global Environmental Change and Human Health: Impact Assessment, Population Vulnerability, and Research Priorities." *Ecosystem Health* 3: 200–210.

Mitchell, G.J. 1984. "The Importance, Utilization, Management and Future of Wild Game Animals on the Canadian Plains." *Prairie Forum* 9: 249–261.

Mosquin, T. and S. Rowe. 2004. "A Manifesto for Earth." *Biodiversity* 5:3-9.

Murdock, S. 1997. "What is Sustainability?" (www.sustainableliving.org, accessed October 6, 2004).

Peltzer, D.A. 2000. "Ecology and Ecosystem Functions of Native Prairie and Tame Grasslands in the Northern Great Plains." *Prairie Forum* 25: 65-82

Radenbaugh, T. A. and P. Douaud (eds.). 2000. *Changing Prairie Landscapes*. Regina, SK: Canadian Plains Research Center.

Rowe, J.S. 1969. "Lighting Fires in Saskatchewan Grasslands." *Canadian Field Naturalist* 83: 317–324.

Samson, F.B. and F.L. Knopf (eds.). 1996. *Prairie Conservation: Preserving North America's Most Endangered Ecosystem*. Washington, DC: Island Press.

Saunders, D.A., R.J. Hobbs and C.R. Margules. 1991. "Biological Consequences of Ecosystem Fragmentation: A Review." *Conservation Biology* 5: 18–32.

Selby, C.J. and M.J. Santry. 1996. *A National Ecological Framework for Canada: Data Model, Database and Programs*. Ottawa/Hull: Agriculture and Agri-Food Canada, Research Branch, Center for Land and Biological Resources Research and Environment Canada, State of the Environment Directorate, Ecozone Analysis Branch.

Sherry, T.W. and R.T. Holmes. 1996. "Winter Habitat Quality, Population Limitation, and Conservation of Neotropical-Nearctic Migrant Birds." *Ecology* 77: 36–48.

Sutter, G. C., and Worts, D. (In press). "Negotiating a Sustainable Path: Museums and Societal Therapy." In R.R. Janes and G.C. Conaty (eds.), *Looking Reality in the Eye: Museums and Social Responsibility*. Calgary, AB: University of Calgary Press.

Tilman, D. 1996. "Biodiversity: Populations Versus Ecosystem Stability." *Ecology* 77: 350–363.

Tilman, D., J. Knops, D. Wedin, P. Reich, M. Ritchie and E. Siemann. 1997. "The Influence of Functional Diversity and Composition on Ecosystem Processes." *Science* 277: 1300–1302.

Venetoulis, J., D. Chazan and C. Gaudet. 2004. *Ecological Footprint of Nations 2004*. Report of the Sustainability Indicators Program. Oakland, CA: Redefining Progress.

Vitousek, P.M., H.A. Mooney, J. Lubchenco and J.M. Melillo. 1997. "Human Domination of the Earth's Ecosystems." *Science* 277: 494–499.

Wackernagel, M. and W. Rees. 1995. *Our Ecological Footprint: Reducing Human Impact on the Earth*. Gabriola Island, BC and Philadelphia, PA: New Society Publishers.

World Commission on Environment and Development. 1987. *Our Common Future*. New York, NY: Oxford University Press.

Wright, H.A. and A.W. Bailey 1980. *Fire Ecology and Prescribed Burning in the Great Plains*. Intermountain Forest and Range Experiment Station, U.S. Department of Agriculture USDA Forest Service General Technical Report INT-77.

Appendix 1:
Ecosystem Service Provided by Various Ecosystem Functions
as Defined and Used by Costanza et al., 1997

Ecosystem Services	Ecosystem Functions
Gas Regulation	Regulation of atmospheric chemical composition.
Climate regulation	Regulation of global temperature, precipitation, and other biologically mediated.
Disturbance regulation	Regulation of global temperature, precipitation, and other biologically mediated fluctuations such as sea level rise.
Water regulation	Regulation of hydrological flows.
Water supply	Storage and retention of water.
Erosion control and sediment retention	Retention of soil within an ecosystem.
Soil formation	Retention of soil and soil formation processes in ecosystems.
Nutrient cycling	Storage, internal cycling, processing, and acquisition of nutrients.
Waste treatment	Recovery of mobile nutrients and removal or breakdown of excess or xenic nutrients and compounds.
Pollination	Movement of floral gametes.
Biological control	Trophic-dynamic regulations of populations.
Refugia	Habitat for resident and transient populations.
Food production	That portion of gross primary production extractable as food.
Raw materials	That portion of gross primary production extractable as raw materials.
Genetic resources	Sources of unique biological materials and products.
Recreation	Providing opportunities for recreational activities.
Cultural	Providing opportunities for non-commercial uses.

CHAPTER 2

Land-Use Policy Change and the Ramifications for Stewardship and Waterfowl Conservation in Saskatchewan

Greg Riemer

A Brief Historical Perspective on Western Canadian Agricultural Policy

In the late 1800s, the Dominion Government in Ottawa was concerned about American annexation of Rupert's Land. The disputed territories of western Canada were a "no man's land" that had to be converted to deeded land and populated to assert Canadian sovereignty. A pattern of privately owned land was established through legislation that was tied to the construction of the railway and the government's ability to grant homesteads. Land tenure became the principal tool of nation building. The original "homestead" was a quarter section of land (160 acres or 65 ha) that was given to any settler for free, provided that he or she lived on it and cultivated a certain portion of it. The railway was completed in 1885 but there was no massive influx of settlers despite hard times and starvation in Europe and massive emigration to the USA.

Twelve years after the completion of the Canadian Pacific Railway's main line, the federal government realized that being more than a thousand miles (1,600 km) from any export position placed grain farming on the Canadian Prairies at a severe disadvantage. The government's response was The Crows Nest Pass Act, which was passed on September 6, 1897. Subsidizing the export of grain made its production on the Canadian Prairies economically viable. This then initiated a land rush that was advertised as the "Last Best West." For the next 100 years, western Canadian grain farmers paid only a portion of the freight bill on exported grain, and the grain industry supported the development of a rural prairie infrastructure. Prairie grain farmers and the industry they created still affectionately refer to the subsidy and the Western Grain Transportation Act (WGTA) that replaced it as the "Crow."

The government's mechanism for securing both western Canada and future prosperity was more people on the land, and the key to keeping them on the land was cultivation. Provisions in federal, and later in provincial, legislation ensured that if land was allowed to "go wild" then homestead rights would be revoked or land taxes increased. These provisions ensured that the new settlers would not allow their land to revert to pasture. Grain production continued to receive increasing government support through a myriad of programs until the early 1990s while the livestock sector was essentially unsubsidized. This resulted in much higher economic rents being paid to land in grain production than in livestock production. The impact of this inequity on the landscape was inescapable, and by the 1960s most of western Canada's good quality soils were cultivated. The trend

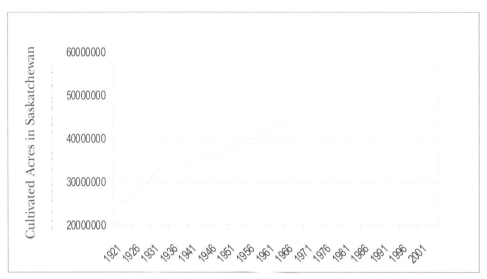

Figure 1. Cultivated acres over time in Saskatchewan. Source: Statistics Canada 1996 and 2001.

toward increased annual crop production continued through the 1980s and is reflected in the land use of the Canadian Prairies today.

The historic, upward trend in cultivation was not constant (Figure 1). A decline in the 1930s was due to drought and depression. A current decline beginning in the early 1990s is likely a result of the removal of grain production subsidies, principally the ending of the WGTA.

The Current Policy Framework

The situation in Western Canada today is much different from the one that dominated most of the previous century. There is recognition that subsidies undermine the profitability of agriculture and have dramatic environmental costs. At the same time, conservation groups and governments have found common ground (e.g., PCAP Partnership, 2003), and agriculture is essentially a deregulated industry. The WGTA that subsidized the rail transport of export grains was removed in the early 1990s. The Gross Revenue Insurance Program (GRIP) that guaranteed an average price for grains was discontinued. The Net Income Stabilization Act (NISA), basically a retirement fund, is now based on whole farms, not just the production of export grains. The Beef Tripartite program (beef price stabilization) was discontinued without making any substantial payouts. The acreage-based quota system which distributed marketing opportunities based on the amount of cultivated land that a farmer owned has been discontinued in favour of a system that divides access to markets based on the farmer's ability to supply grain on contract. The Western Canadian Wheat Board remains, but it is not really a subsidy. Support for it is mixed, and its future is uncertain.

Annual ad hoc acreage-based subsidy programs (such as drought payments) have not been used in many years. This has strengthened the need and demand for crop insurance, which is the only subsidy that remains. To the Crop Insurance Program's credit, they have recently introduced forage and pasture insurance programs that have removed any bias in favour of grain production. Crop insurance programs will be forced to undergo major changes as (or if) the regulations under the General Agreement on Tariffs and Trade (GATT) come into effect. The international response will likely be to move towards a much shorter yield-averaging period, which would cut the peaks and valleys off the yield cycle and would likely reduce the value of crop insurance to farmers over the long term.

Both the federal and provincial governments are aware that pro-grain production policies have resulted in the cultivation of much land that is physically marginal for grain production. Over ten years ago, the federal Permanent Cover Program converted almost a million acres (over 400,000 ha) to forage nationwide. Four years ago the government of Saskatchewan announced a Conversion Cover Program with a $5 million annual budget. It offered a $15 per acre ($37/ha) one-time payment to assist with seeding marginal cropland back to permanent cover, with no restriction on land use. This program was oversubscribed every year and is no longer available. Recently, the federal government announced a "Green Cover Program" to assist with the conversion of margin cropland back to grass.

Recent Landscape Changes

Recent census data show a decrease in the amount of farmland in Saskatchewan, an increase the amount of tame hay, and a decrease in the amount of cultivated land over the last two census periods (Statistics Canada, 1996, 2001). The area defined as farmland declined by 1.1% in both 1996 and 2001, reducing farm area by 2.2% in just 10 years. This amounts to 1.43 million acres (0.6 million ha) that are no longer part of active farming operations.

The acreage of cultivated land in Saskatchewan can be adjusted for tame hay production by comparing agriculture census data (Statistics Canada, 1996, 2001) to provincial statistics (Saskatchewan Department of Agriculture and Food, 2000). The results show that the amount of cultivated land in the province, adjusted for tame hay production, has declined from all-time highs in the late 1980s and early 1990s to a level in 2001 that was only 0.5 million acres (0.2 million ha) higher than cultivated acreages of the 1960s and 1970s (Figure 2). In other words, there was only a 1% increase in the amount of land in annual crop production between the early 1970s and 2001 (Figure 2). This is a dramatic recovery, given that almost 3.5 million acres (1.4 million ha) of permanent cover were lost due to increased cultivation between the 1970s to the 1980s (Figure 1). The increase in tame hay production was so significant that tame hay went from being a minor crop on older crop acreage reports to the fourth largest crop in Saskatchewan in the 2001 Agricultural Census.

The trend towards less cultivated land is continuing. It is being driven by poor grain profitability and the continuation of American and European grain production and export subsidies. In 2002, the Saskatchewan Conversion Cover Program seeded 296,244 acres (almost 120,000 ha)[1] to permanent cover, even with a limit of 50 acres (20 ha) per farmer. Their clients noted that they had seeded an additional 101,769 acres (over 42,000 ha) for which they were not eligible for payment. This amounts to 398,013 acres (162,000 ha) of cover seeded in 2002 that is not shown in Figure 2. It also means that the amount of land under permanent cover is likely higher now than it was in the early 1960s, assuming that no existing tame hay has been cultivated. Farmers who did not apply for the one-time subsidy are also not included in this data.

Figure 2. Saskatchewan acres in cultivation adjusted for hay production. Source: Statistics Canada, 1996 and 2001.

Changes in the Landscape Resulting from Technological Change

The reduction in the total amount of land farmed and the permanent cover increases are important from a conservation perspective. A major driving force on the landscape of Saskatchewan, and the rest of the northern Great Plains, has been the development of effective zero till or minimum tillage systems using modern air seeders. Seeding is now fast and efficient with minimal disturbance of the soil, substantial crop carry over or "trash," and significant savings in diesel fuel and capital costs. As a result, the technology has been adopted quickly.

The first casualty of this technology was summer fallow, a practise that originally enabled the prairies to be farmed and was long considered the "right" way to manage farmland. In Saskatchewan, the amount of land in summer fallow was relatively stable in the 1970s at slightly more than 17 million acres (6.9 million ha). With high carryovers of wheat in the early 1970s, the federal government actually paid farmers to summer fallow in 1971 with the Lower Inventories For Tomorrow program (LIFT). Summer fallow coverage peaked in Saskatchewan at a high of approximately 24 million acres (9.7 million ha)[2] and fell by almost 50% from 17 million acres (6.9 million ha) in the 1970s to 8.4 million acres (3.4 million ha) in 2000 (Figure 3).

The removal of agricultural grain production subsidies has changed the economic picture so much that many producers are moving into forage-based agriculture, and this trend is environmentally positive. However, unlike changes to permanent cover that affected several million acres, continuous cropping has impacted 10s of millions of acres of cropland. Over time, as summer fallow disappeared, half or two-thirds of summer fallow-cereal rotations have been converted to minimum tillage. Population estimates suggest that the adoption of continuous cropping practises has been beneficial or benign for most waterfowl. Aside from the Northern Pintail (*Anas acuta*), the conservation of waterfowl populations in prairie Canada appears to be a "just add water" problem.

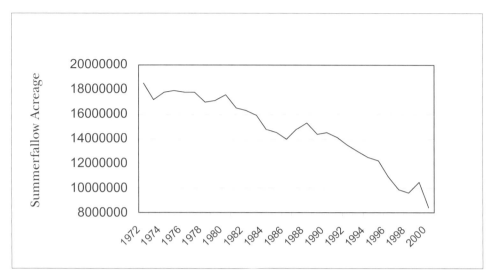

Figure 3. Summer fallow acreage in Saskatchewan. Source: Saskatchewan Agriculture and Food, 2000.

The Importance of Saskatchewan to Waterfowl Conservation

The agricultural region of Saskatchewan is a mosaic of cultivated fields and pastures, uplands and wetlands that overlay a highly varied topography ranging from flat glacial lakebeds to very hilly and hummocky glacial moraines. These moraines, created by retreating glaciers, are part of the North American prairie pothole region that contains approximately 25 million wetlands and is the principal nesting ground for most of North America's waterfowl. Saskatchewan comprises a major share of the wetlands, more than any other province or state, so changes in Saskatchewan's agricultural landscape will have large and direct effects on many grassland-nesting birds such as songbirds, waterfowl, shorebirds, and upland game birds.

Based on statistics collected since the 1950s (US Fish and Wildlife, 2002), waterfowl populations are closely correlated with spring pond numbers on the northern Great Plains (Figure 4), with high populations corresponding to good moisture conditions in the 1950s and 1970s (Figure 5). By the late 1980s, after a period of prolonged drought, waterfowl numbers had crashed and alarm bells began to ring within the waterfowl conservation community.

In a paper that helped launch the NAWMP, Johnson and Shaffer (1987) noted that mallard populations were lower from 1971 to 1985 than the spring pond count would have historically predicted. They hypothesized that mallards were no longer fully using their habitat. Their data, and later reports from Statistics Canada (1996, 2001), showed that habitat conditions in Saskatchewan had changed. From the early 1970s to the late 1980s, almost 3.5 million acres (1.4 million ha) of native prairie in Saskatchewan had been converted to wheat fields. After much scientific, economic, and political work and negotiations, the NAWMP was born. Waterfowl numbers rebounded in Saskatchewan and elsewhere as moisture conditions improved through the 1990s (Figure 4), with continental populations reaching historic highs in 1999 (Figure 5).

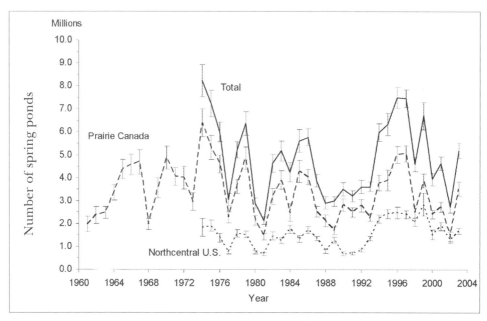

Figure 4. Spring pond counts in the prairie pothole regions of North America. Source: US Fish and Wildlife Service, 2002.

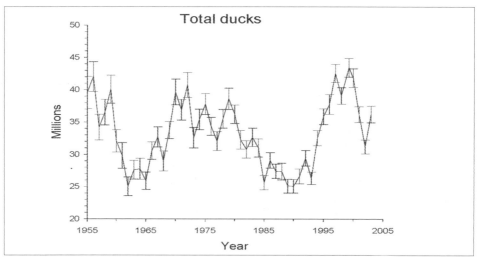

Figure 5. Total ducks on North American flyways. Source: US Fish and Wildlife Service, 2002.

The Impact of Landscape Changes on Waterfowl

The end result of more than one hundred years of Canada's cultivation-based settlement policy is one of the most altered landscapes on the planet. In Saskatchewan, 65 million acres (26 million ha) had been brought into agriculture by the late 1980s, and roughly 75% of this land was cultivated (Figure 1). The increase in cultivation in the late 1970s and 1980s had a particularly negative impact on waterfowl populations. Almost all of Saskatchewan's Class 1, 2 and 3 farmland was cultivated[3] by the early 1950s, so most of the land that has been cultivated since is marginal for grain production. Many researchers believe that this loss of habitat caused a major downturn in waterfowl numbers.

Some analysts have suggested that hay land may be less beneficial for birds than cropland if it attracts species that are subsequently killed during haying operations, but this hypothesis is not supported by the recent increases in populations of late-nesting waterfowl. As indicated earlier, the number of total waterfowl in North America peaked at all-time highs with the end of the wet years in 1999 (Figure 5). Massive habitat changes associated with the conversion of marginal farm land to permanent cover, including millions of acres in the U.S. Conservation Reserve Program (CRP), and the advent of zero tillage accompanied by a reduction in summer fallow, appear to have been very beneficial to waterfowl and other birds.

However, not all waterfowl species have reacted to habitat changes in the same way. The Mallard (*Anas platyrhynchos*), often considered the most important species from a hunting and recreational perspective, has done very well, as their numbers increased and decreased following the cycle of spring pond counts from the early 1950s to the late 1990s (Figure 6). This was not the case for late-nesting ducks such as the Gadwall (*A. strepera*), Northern Shoveler (*A. clypeata*), Green-winged Teal (*A. crecca*) and Redhead (*Aythya americana*). For example, instead of following the spring pond count, Gadwall had low but relatively stable populations until the 1990s, when their numbers increased dramatically and then fell off as pond numbers declined after 1999 (Figure 6). Why some waterfowl populations were not linked to spring pond counts merits further research. It may be that later-nesting ducks have benefited from upland habitat changes on the prairies.

Problems with the Northern Pintail

The one prairie-nesting duck that has not responded favourably to habitat changes is the Northern Pintail. The Pintail's North American population has been steadily declining through repeated wet cycles, even through the most recent one (Figure 6). The lack of rebound in Pintail numbers in the 1970s was probably a result of the cultivation of native prairie that occurred from the 1950s to the 1970s (Figure 1). However, if available cover and pond numbers alone were responsible, then the levelling of the amount of permanent cover from the 1970s through the 1990s should have allowed for a rebound to at least the level of the 1970s. Something else besides the amount of permanent cover on the landscape is impacting Northern Pintail populations.

The Temporal Significance of Summer Fallow for Prairie Nesting Ducks

To understand how waterfowl have been affected by a reduction in the amount of summer fallow and its converse, the increase in minimum tillage, we need to consider how the timing of cultivation affects the breeding and nesting of various species. A summer-fallowed field is generally barren and dark in colour, but it does not reach that condition until late spring or early summer the year after the crop was combined. It may or may not have been worked once after harvest in the fall, depending on moisture conditions, and it is always left untouched in the spring until the previous year's summer fallow is seeded in May and early June. In years with adverse farming weather, it may be cultivated for the first time as late as mid- to late June, after the completion of spraying for weed control. In most years, until

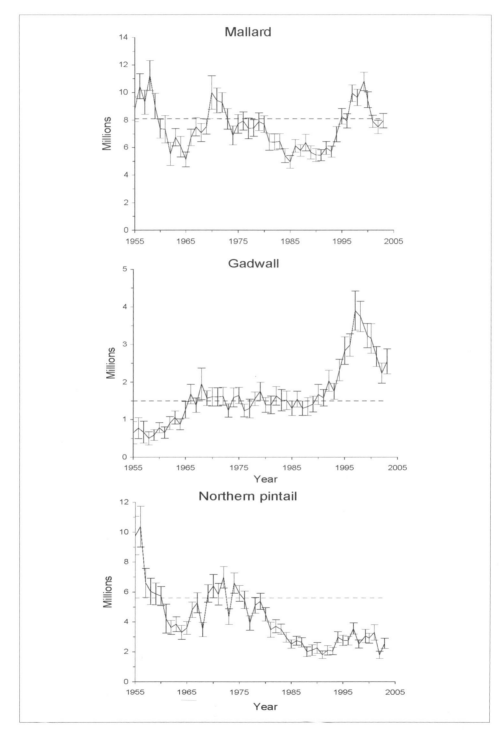

Figure 6. Mallard, Gadwall, and Northern Pintail populations over time in North America. The dotted line is the NAWMP population objective for each species. Source: US Fish and Wildlife Service, 2002.

early June, it is attractive habitat for early-nesting birds that utilize low cover, like the Northern Pintail. Summer fallow from the previous year that is seeded in May will be a barren field of dirt that would have low attractiveness, even to Pintails.

Important differences are apparent when the practice of summer fallow is juxtaposed with continuous cropping. When summer fallow is practiced in a half-and-half rotation, half the land is cropped and half is tilled. With minimum tillage, all the land is cropped but it is never reduced to bare soil and it is almost never worked in the fall, which allows a substantial amount of dead vegetation, or litter, to build up, a habitat that attracts nesting Pintails. The problem is that minimum tillage fields are disturbed by seeding operations in May, just as (or shortly after) Pintails start nesting. Many hens and their offspring are killed, and the reproductive output of the population is reduced because Pintails are reluctant to renest.

Guyn et al. (2002) postulate that nesting Pintails take advantage of cropland still present as unworked stubble and may get their broods off of the fields prior to the first June tillage. They also observe that the Pintail population's lack of rebound when moisture increased in the 1990s appears to follow the reduction of summer fallow (in actuality, the adoption of continuous cropping) during the early 1980s and 1990s. As stated earlier, if pond counts and available cover were the only limiting factors, Pintail numbers should have rebounded to the level of the 1970s in the late 1990s.

The opposite situation exists for late-nesting or re-nesting ducks. Once seeding is completed in May or early June, the only impacts on the field are due to wide sprayers that are unlikely to damage nests, and most waterfowl appear to find the cover provided by the crop in June more attractive than stubble. Not surprisingly, populations of many species of prairie-nesting waterfowl like the Gadwall, Blue-winged Teal, and Northern Shoveler, which were relatively stable through past moisture cycles, expanded spectacularly in the 1990s (US Fish and Wildlife Service, 2002).

Continuous cropping is clearly not the only the reason for the rebound of many species of waterfowl in the 1990s. Other land use changes, such as the conversion of marginal cropland to permanent cover in Canada and the American CRP, have likely had many beneficial impacts. However, only continuous cropping can account for the ongoing decline in Pintail numbers. With the adoption of continuous cropping over the last 30 years, over half of the cropland base (approximately 30 million acres in Saskatchewan alone) that offered suitable nesting habitat for this species may have been converted into an ecological sink. Seeding marginal cropland back to pasture may be the best way to remove this threat. However, even if all of the marginal cropland in the northern Great Plains were converted to pasture, the remaining continuously cropped land would still be a potential trap for these birds. Studies of Pintail nesting ecology and habitat use, especially the possibility that spring sheet water on cropland may be encouraging Pintails to nest in cropland, should be a research priority if we hope to restore Pintail populations to historic levels.

Impact of the NAWMP on the Western Canadian Landscape

The agricultural landscape of Saskatchewan is clearly changing at a rate that cannot be accounted for by NAWMP activities. NAWMP census evaluations through the Prairie Habitat Joint Venture (PHJV) Habitat Monitoring Program imply that

Table 1: Agricultural Census Data: Upland Habitat Changes from 1986 to 1996

Cover Type	NAWMP Target Area			Remaining Non-Target Landscape		
	1986	1996	Change	1986	1996	Change
Summer Fallow	14.53%	10.53%	- 4.0%	15.99%	12.33%	- 3.66%
Annual Crop	45.68%	46.93%	+ 1.25%	44.78%	45.69%	+ .91%
Total Cultivated	60.21%	57.46%	- 2.75%	60.77%	58.02%	- 2.75%
Native Pasture	26.13%	25.21%	- .92%	23.88%	23.36%	- .52%
Tame Pasture	4.3%	6.33%	+ 2.03%	4.57%	6.14%	+ 1.57%
Tame Hay	3.79%	5.15%	+ 1.36%	5.11%	6.79%	+ 1.68%
% Cover Change			+ 2.47%			+ 2.73%
All Other land	5.58%	5.84%	+ .26%	5.67%	5.68%	+ .01%

Note: Total cultivated land is the sum of summer fallow and annual crop, and % cover change is the sum of changes in native pasture, tame pasture, and tame hay.

land is being converted into forage-based agriculture at almost the same rates in NAWMP target areas as in non-target areas. The program's first report by Watmough et al. (2002) describes wetland and upland landscape changes on the ground in NAWMP target and not-target areas. Based on samples from 152 transects, each containing 24 quarter sections for a total of 583,680 acres (over 236,000 ha), they observe that the trend to forage-based agriculture is large and more prevalent in NAWMP target areas than on the general agricultural landscape.

Watmough et al. (2002) also compare Statistics Canada Agricultural Census data from NAWMP target and non-target areas. From 1986 to 1996, differences in the rate of change were very small between the two landscapes (Table 1). The differences that did exist were most likely a result of the fact that the NAWMP target areas are glacial moraine landscapes that contain high wetland densities and are generally less suitable for the production of grain. The NAWMP target areas had less summer fallow but more annual crop after 10 years, but there was no difference in the rate of change in total cultivated acres between target and non-target areas. In addition the NAWMP targets had lost more native prairie and gained tame pasture over that period, but they also showed a smaller increase in tame hay compared to the non-target landscape. Overall, the rate of increase in permanent cover was slower in NAWMP targets than in non-target areas. Moreover, at the end of 10 years, the difference in the rate of change in permanent cover on the two landscapes is almost entirely accounted for by the 0.26% increase in the "All Other Land" category of the target areas (Table 1).

The percentages noted above are for all of western Canada, but they closely parallel the Saskatchewan numbers this paper uses to adjust the amount of cultivated land for hay production. Based on provincial numbers (Saskatchewan Department of Agriculture and Food, 2000), tame hay coverage in Saskatchewan increased from 1.86 million acres (0.75 million ha) in 1986 to 2.95 million acres (1.19 million ha) in 1996. This represents a change to 1.68% of the 65 million acres (26 million ha) of farmland in the province, which is the rate of increase observed outside NAWMP target areas (Table 1).

Conclusion

The reduction in the amount of land farmed, increases in the amount of permanent cover, especially in recent years, and the adoption of continuous cropping on 10s of millions of acres have dramatically impacted the Saskatchewan agricultural landscape. Individual farmers and ranchers, who are responding to economic pressures, are changing the landscape at rates that are almost completely masking the impact of the NAWMP. As the profitability of crop production has declined, the abandonment of farmland has become a serious concern to rural economies and almost twenty-times more land has been affected than conservation agencies have the ability to buy, even if they wanted to.

In terms of waterfowl conservation, much of the prairie landscape has changed for the better for most species not only in Canada through subsidy removal but also in the United States through the American Farm Bill. Waterfowl have responded, and in the late 1990s, demonstrated that the habitat is there for record waterfowl production on a continental basis. Our attention must now be directed to species such as the Northern Pintail that are still at risk or appear to be impacted negatively by these landscape changes.

Based on these observations, the following recommendations are offered:

• The impacts of agricultural deregulation in Canada and large-scale environmental programs in the USA have had dramatic impacts on land use. Policy reform activities have produced the most cost-effective landscape changes and should be continued.

• Migratory bird plans should be revised to account for changes at the landscape level in the USA and Canada and habitat trends on the northern Great Plains in general.

• Migratory bird plans should use long-term (decadal) population trends rather than short-term (seasonal) trends as indicators of population health. The definition of what constitutes a stable population needs to be delineated so that more useful comparisons can be made.

• Migratory bird plans should remove bias towards the importance of a species in the harvest and emphasize species that appear to be responsive to increased moisture and anthropogenic landscape changes.

• With good moisture conditions, waterfowl populations are now able to reach record highs. Thus, American and Canadian federal money should be targeted at the broader North American Bird Conservation Initiative, not just towards waterfowl.

• Special attention should be given to landscape problems associated with reductions in the amount of summer fallow and increases in continuous cropping and the influence of these changes on Northern Pintail. Likewise, the reasons for rapid increases in populations of late-nesting waterfowl need to be examined.

• Reducing the attractiveness of continuously cropped land to early nesting waterfowl should be a priority. Specifically, the impacts of sheet water on continuous cropped land on the breeding, nest-site selection and nesting success of these birds needs to be investigated.

Acknowledgements

Earlier versions of this manuscript were greatly improved by input from Glenn Sutter and two anonymous reviewers.

Notes

1. Personal conversation with Lee Giroux, Manager, Saskatchewan Conservation Cover Program.
2. Saskatchewan Department of Agriculture and Food, *Saskatchewan Agriculture and Food 2000* (Regina: Saskatchewan Department of Agriculture and Food, 2000), 28.
3. According to Statistics Canada, cultivated land is any land in annual crop, hay, or seeded pasture less than five years old. From a waterfowl production perspective, land in hay or pasture is dramatically different from land in a wheat-fallow rotation.

References

Agricultural Statistics 2000. *Saskatchewan Agriculture and Food 2000.*

Guyn, K.A., J.H. DeVries, K.M Podruzny and L.M. Armstrong. 2002. "The Deception of Pintails on the Canadian Prairies." In *Proceedings of The Wildlife Society 9th Annual Conference, 2002.* Bismarck, ND.

Johnson, D.H. and T.L. Shaffer. 1987. "Are Mallards Declining in North America?" *Wildlife Society Bulletin* 15: 340–345.

PCAP Partnership. 2003. *Saskatchewan Prairie Conservation Action Plan 2003–2008.* Regina: Canadian Plains Research Center.

PHJV Implementation Strategy. 1989. Prairie Habitat Joint Venture Committee. Canadian Wildlife Service, Environment Canada. Edmonton.

Statistics Canada. 1996. Agriculture Census Data 1996 www.statcan.ca

——. 2001. Agriculture Census Data 2001 www.statcan.ca

US Fish and Wildlife Service. 2002, *Waterfowl Population Status, 2002.* Washington, DC: US Department of Interior.

——. 2003, *Waterfowl Population Status, 2003.* Washington, DC: U.S. Department of Interior.

Watmough, M.D., D.W. Ingstrup, D.C. Duncan and H.J. Schinke. February 2002. *Prairie Habitat Joint Venture Habitat Monitoring Program Phase 1: Recent Habitat Trends in NAWMP Targeted Landscapes.* Environment Canada.

CHAPTER 3

Energy Trends For Saskatchewan Farming, 1936–1991

Bob Stirling

Introduction

Rising oil prices in the 1970s left downstream industries worrying about their viability and revived concern over the efficient use of energy. Agriculture was no exception. For example, at a 1977 seminar on "Energy Conservation in Agriculture" sponsored by Saskatchewan Agriculture, Mathieson (1977: 49) estimates that if fuel prices continued to rise so that by 1981, gasoline was $1.05 per gallon and diesel was $0.93 per gallon, then "wheat prices would have to average in excess of $6 per bushel to maintain the long run average margin of motive fuel costs to gross return per seeded acre." The current price for fuel far exceeds those 1981 estimates while the price for wheat has languished. Not surprisingly, governments and industries have sponsored studies to inform their reactions to the impending crisis.[1]

There are, of course, other criteria for evaluating farm policies besides energy use and the energy efficiencies of farm production. Policies concerning European farm settlement in the Canadian Prairies were originally judged in terms of maximizing the output of an export staple, and creating a home market for industrial commodities (Fowke, 1957). Recently, governments have turned their attention to "sustainability." Under this rubric, Canadian farm policies have been judged in terms of their ability to sustain a farm labour force and farm families with decent income (e.g., Government of Canada, 1969), as well as to sustain the natural resources needed for farm production such as the quality of soil, water, and air, as well as agroecosystem biodiversity (e.g., McRae et al., 2000). While declining energy efficiency implies a farming system that is shedding labour, the ratio of energy outputs to energy inputs is probably not a good proxy for the sustainability of farming's social environment since the market often disciplines farmers to overlook the energy implications of their production decisions (Stirling, 1979). Still, it may be a useful summary measure of the sustainability of farming's physical environment since greater output energy per unit of input energy implies a farming system that either uses less fossil fuel and industrial inputs or else uses them to greater effect. As Moseley and Jordan (2001) point out:

> Viewed from the energy efficiency perspective, sustainability is a relative concept, and the system that produces the most from the least could be considered to be the most sustainable.(pp. 2–3)

Hence, a worthy task has been to better understand the energy efficiency of northern prairie agriculture by quantifying the farm energy costs of our food supply compared to the food energy that it contains.[2] This study addressed production on the farm, supplying commodities "to the farm gate."[3] Downing and Feldman

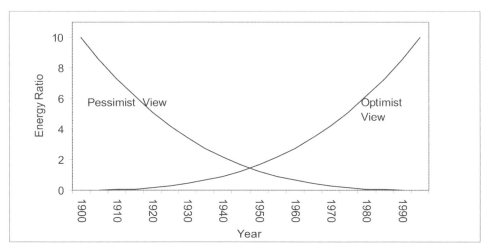

Figure 1. Alternate Views of the Trend in Farm Energy Efficiency.

(n.d.) had estimated the output-input energy ratio for Canadian agriculture to be quite high, perhaps because they were unable to estimate the energy volumes for the full range of farm inputs. Some estimates for Canadian prairie wheat had also put the energy ratio as high as 8.7:1 (Slesser, quoted in Hulse, 1995: 145) whereas ratios between 4 to 5 seemed more likely (Timbers, 1977). Ratios for animal production are typically below 1 to 1 (Pimentel, 1980; Southwell and Rothwell, 1977; Stirling, 1979). Ratios for the prairie region's total compliment of farm production were difficult to estimate since most of the studies had focussed upon specific commodities.

Equally important is the trajectory of agriculture's energy efficiency about which there are competing views (Figure 1). An "optimist" view, which was widely held within agricultural circles after World War II, saw agriculture becoming ever more efficient as it adopted science-driven technologies. For example, reflecting what would now be seen as a progressivist approach, a major critical inquiry into agrarian life in the 1950s worried about things that inhibited farmers from raising productivity by adopting technology (Saskatchewan Royal Commission on Agriculture and Rural Life, 1957).

A "pessimist" view has become popular since the initial rise of petroleum prices in the 1970s and with a growing concern about agricultural sustainability. For some, declining farm energy efficiency is only one of the many ways that humans, perhaps through ignorance and misunderstanding, but also wilfulness, have been doing violence to the environment and other living things (e.g., Rowe, 1990). Others have shown that as agriculture has become more industrial and "modern" it has become less efficient in its use of energy. For example, Pimentel and Burgess (1980: 67–75) present estimates of output: input energy ratios (Er) for several forms of corn production as follows: hand-powered (Mexico), Er = 128.2; Oxen-powered (Mexico), Er = 4.25; and modern (North Dakota), Er = 1.98. Leach reports on changes in Er during the UK's "full-industrialised" agriculture period. The estimated energy ratio dropped from 0.46 in 1952 to 0.35 in 1972. Simply put, "while UK agriculture became more efficient in many ways it became much less efficient in its use of energy" 1976: 24).

Our earlier work was inconclusive on the question of an historical trend (Stirling, 1979). For the period 1961–76, we found that the Er trend for Saskatchewan farming tended to be flat or slightly downward with the exception of a dry period in 1961. Perhaps this trend for 1961–76 was the bottom of a longer term declining curve. Alternately, perhaps it was the top of what had been an increasing curve.

Hence, two questions were the impetus for this work. First, what is the historical pattern for energy efficiency of farming in Saskatchewan, that is, has the energy ratio tended to remain steady, to increase, or to decrease? Second, can we learn something from the historical pattern that might be relevant for farm policies? If Er has been improving then the current direction of farming and its related policy configuration would appear to be appropriate from the single criterion of energy efficiency. If Er is decreasing then the current direction of farming and its related policies are called into question, as far as energy efficiency is concerned. If the output-input ratio has not changed appreciably, then the promise of technology-inspired farming may only partly be realised; greater inputs have yielded greater outputs but, perhaps, at some risk to the physical and social environments.

Farm Structure

There have been compelling reasons for prairie farmers to increase energy inputs and outputs. On the one hand, farms are economic units that must respond in some manner to markets for inputs and outputs. On the other hand, the evolution of prairie farming has been modified by other influences. Prairie farms have been run by families, with family considerations sometimes dampening and other times heightening market imperatives. The evolution of the farm sector has also been shaped by the Canadian Prairies as a distinct geographic region.[4]

The basic structure of Saskatchewan agriculture was originally set by federal policy, specifically, the "national policy" and related legislation put in place between Confederation and World War I (Fowke, 1957). This had the effect of closely tying prairie farms to the imperatives of domestic and international markets. Prairie agriculture supplied export staples, particularly, wheat and other grains. This led to a specialised farming system especially apparent in Saskatchewan where crop production accounted for 95% to 98% of its total value of farm production.

For the "national policy" to be successful, prairie farmers were encouraged to adapt their specialised agriculture to the dry conditions and shallow soils of the northern plains and the onslaught of annual, "broad-leaf" weeds. The practice of summerfallowing became common. It, in turn, was enhanced by the invention of tillage machinery and gasoline tractor power. The weeds were ultimately controlled by combinations of tillage and chemicals (particularly 2,4-D after World War II) but these were followed by hardier annual and perennial weeds which led to a new round of technology to combat them (e.g., wild oat herbicides and glyphosate). The persistent need to boost productivity encouraged the invention and use of new chemical fertilisers. Many of these technological "remedies" migrated north with American companies seeing the Canadian Prairies as a natural extension of their market (Shepard, 1994). The need to produce staples for the international market, the growing conditions of the northern plains region, and the influence of American agri-business have enhanced the enthusiasm of prairie farmers towards adopting technology.

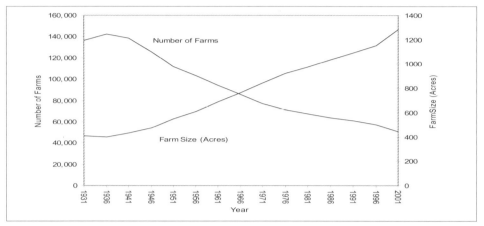

Figure 2. Number of Farms, and Farm Size, Saskatchewan, 1931–2001.

The structure of agriculture set by the "national policy" included many thousands of family farms. The farms and farm families created a market for manufactured goods and services, with farmers becoming increasingly dependent upon non-farm firms for inputs. Farmers have faced a cost-price squeeze which has arguably been more serious for them than for other industries. Similarly, the margins for Saskatchewan farming have often been low and mostly declined after 1975. These two features have impelled farmers to adapt, typically by adopting technologies from non-farm firms.

Concentration of farm capital and declining farm numbers have been some other inherent characteristics of this structure. Farm families have responded to tight margins in a variety of ways, such as working harder, accepting lower incomes, poorer living conditions, undercapitalised farms, and sending some members to take off-farm work. But, overwhelmingly, farm families have shed labour to urban areas and industries. From 1931 to 1991, Saskatchewan lost about 400,000 farm people and over 75,000 farms. Average farm size (in area) grew over 250%. The sharpest change came during and immediately after World War II (Figure 2). Since farm production generally did not decline, some portion of farm labour became expendable only when it could be replaced with non-labour inputs, such as bigger machinery, more chemicals, and more fertiliser. In other words, the structure encouraged farmers to substitute technology supplied by non-farm firms for farm labour, a tendency shared with farming systems historically and in other regions (Leach, 1976; Marchetti, 1979).

The sequence by which the major technologies were adopted is well known (Table 1). European farming on the Canadian Prairies began with oxen and

Table 1. Sequence of Some Major Technological Adoptions in Prairie Farming.	
1910s+	Steam tractor power
World War I+	Summerfallowing
1920s+	Gas tractor and related machinery
World War II+	Electricity, diesel, fertiliser, pesticide
1970s+	No-till, low-till
1980s+	Computerisation, new chemicals
2000s+	GM crops and animals

horses, but the steam tractor is credited with breaking much of the native sod. The Dryland Farming Movement introduced shallow tillage and summerfallowing that became universal after World War I. Initially, most of its machinery was horse drawn. Although gas tractors were available in the 1920s, tractor power did not seriously invade prairie farming until World War II. The Saskatchewan horse population reached its peak in 1926 at 1,104,300. It still hovered around 800,000 in the early years of World War II, but then dropped to under 100,000 and by the early 1960s, it was less than 10% of its 1920s population. If we conservatively estimate the decline of working horses at about 1 million, and use the methodology of the Saskatchewan Royal Commission on Agriculture and Rural Life (1955a, Appendix VI and pp. 34–35), it appears that at least 10.5 million acres (4.25 million ha) were released for the production of commodities other than feed for the horses when tractors replaced horse power in the province. Of this total, about 5.5 million acres (2.23 million ha) were crop land or about 12% of the crop land available in 2001. Another 5 million acres (2.02 million ha) of pasture would have been released, or 54.8% of the total pasture and unimproved land in 2001.[5] Depending upon changes in the composition of farm products, this historic transition from horse to tractor power could have added some 10–12% to the outputs side of the energy ratio.

On the inputs side, the trends show the important influence of World War II. Whereas in 1941 only 39% of farms had a tractor, by 1951 95% did, and in 1981 there were, on average, 2.15 tractors per farm. Similarly, while electricity was available much earlier, it took the development of a central power utility to increase electricity use on farms (White, 1976). The results were dramatic. For example, the proportion of farms with power tools increased from 42% to 59% in just three years (1950–53), while the proportion with electrical water pumps increased from 38% to 46% (Royal Commission, 1955b). Many of the modern herbicides were war inventions turned to farm use. During the 1960s, farmers sprayed between 40% to 60% of Saskatchewan's land in crops with either 2,4-D or MCPA (estimated using data from Saskatchewan Agriculture and Food, selected years). New generations of herbicides such as the wild oat herbicides and glyphosate have also seen rapid adoption. Finally, fertiliser sales, while variable, roughly doubled from the end of World War II to the early 1950s, then rose some 10-fold by the mid-1970s and 25-fold by 2000 (Statistics Canada, #46-207; Saskatchewan Agriculture and Food, selected years).

After the high yields in the 1920s that were associated with the fertility of the newly cropped prairie soils, yields tumbled in the drought of the 1930s but have gradually increased since World War II. Along with the rise in inputs went the continuous development of new varieties of crops and, more recently, a trend away from summerfallow and toward continuous cropping and reduced tillage. McGregor et.al. (2000, p. 175) report that energy from crops on the prairies grew by 1.5% annually between 1981–85 and 1992–96, which they attribute to increasing yields and more seeded acres as a result of less summerfallow. Similarly, some livestock have seen improvements in feed efficiency. The question is whether the many socio- economic and technological changes have changed the energy efficiency of Saskatchewan farming.

Study Methodology

Two techniques were used to address this question. First, data collected by public agencies such as Statistics Canada and Saskatchewan Agriculture and Food afforded a window of opportunity to build an inventory of estimated energy inputs and outputs for Saskatchewan farms for 1961–91. Second, these estimates were extended back to 1936 using expenditure data for inputs and volume indexes for outputs.[6]

The energy inventory is simply an accounting system of all of the non-renewable energy embedded in farm inputs and outputs. This includes direct energy such as in farm fuels and lubricants and in electricity (over the period of study most of Saskatchewan's electricity came from non-renewable sources). It also includes the indirect energy for inputs such as machinery, fertilisers and chemicals, building supplies, and commercial feeds and seeds that was embedded in them during their manufacture. Building the inventory involved finding counts of the amounts of each input and output commodity then multiplying these amounts by estimates of the quantity of energy embedded in each of them. For example, on the inputs side, the number of tractors in the 60 to 90 horsepower range was multiplied by an estimate of the amount of energy embedded in this size of tractor (in its steel, rubber, and other components). On the outputs side, for example, the amount of wheat in pounds was multiplied by an estimate of the metabolisable or edible energy in each pound. Several standard references were used for these estimates including Leach (1976), Pimentel (1980), Pimentel et al. (1973), Southwell and Rothwell (1977), Jensen (1977), and others. (For a more detailed description of this technique and estimates, see Stirling, 1979; Stirling and Kun, 1992).[7]

Although this energy estimation technique is useful, it has several limitations. It depends upon publicly available data that may have missing values, for example, in order to protect manufacturing firms. This was especially problematic beginning in the 1980s for pesticide data and some machinery data. Estimates based upon Census and other public data were used to fill in the missing cells. In addition, the energy efficiencies of some types of manufacturing, mining, smelting, and refining have improved, so the coefficients used here may underestimate the pre-World War II energy use.

In order to extend the energy estimates based upon the inventory method back to the Great Depression era, a slightly different approach was used for inputs and outputs. On the inputs side, time series for the seven categories of farm expenditures in the inventory, that is, machinery, fertilisers, pesticides, electricity, buildings, seed and feed, and fuel were converted to constant dollars. Using the inventory estimates, a coefficient representing the cost of embedded energy for each input category was calculated then this coefficient was applied to the actual input expenditures for the years prior to 1961. The process of determining the coefficients included calculating the average cost of energy over the period of inventory values, that is, 1961 to 1991. Where it was clear that the type of inputs in a category changed significantly toward the end of the 1961–91 period, the earlier years were weighted more heavily in estimating the cost coefficient. For example, the pesticides used in the 1960s (2,4-D primarily) are more similar to those used in earlier years than are the wild oat herbicides and glyphosate used in the 1980s, hence the estimate was adjusted accordingly. The intention was also to allow the two series to join more seamlessly from the late 1950s to the early 1960s.[8] On the outputs side, time series for the index of production are available only for total production and, after 1950, for two categories ("crops" and "animals and animal products"). The inventory of output energy was aggregated into these two categories and the indexes used to project output energy for the years prior to 1961.

Clearly this aggregation technique has limitations. For one thing, the categories for which time series are available may be too broad, or put another way, the basket of commodities in each category is bound to change somewhat over time. While available time series data on the inputs side matched the categories in the energy inventory for the period 1961 to 1991, on the outputs side they required aggregation into two categories ("crops" and "animals and animal products") from 1960 back to 1950, and then into one category ("total farm production") prior to 1950. However, considering the specialisation of Saskatchewan agriculture and the dominance of crops over livestock production, this may not be a serious limitation, especially for this early period when wheat was "king" in Saskatchewan farm production.

Study Results

Both farm input and output energy rose on Saskatchewan farms during the 55-year study period (Figures 3 to 7).[9] The energy attributable to both farm outputs and farm inputs increased. In addition, farm output energy showed considerable variation, no doubt reflecting the vulnerability of farm production to cycles of weather and markets. However, on balance, for the 1936 to 1991 period, the energy ratio (energy outputs divided by energy inputs) appears to have been flat to slightly negative. The components of this trend are described below.

Farm Input Energy

Table 2 and Figure 3 show the components of farm input energy in billions (or 10^{12}, or American trillions) of kilocalories. Over the period, total input energy

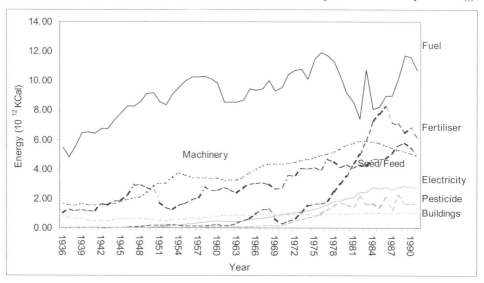

Figure 3. Components of Farm Input Energy, 1936–1991.

Table 2. Components of Farm Input Energy, Saskatchewan, 1936 to 1991 (in 10^{12} kcal).

Year	Machinery	Fertiliser	Pesticides	Electricity	Buildings	Seed/Feed	Fuel	Total
1936	1.67	0.03	0.02	0.00	0.79	1.01	5.46	8.98
1937	1.56	0.02	0.03	0.00	0.70	1.26	4.81	8.39
1938	1.59	0.03	0.03	0.00	0.67	1.18	5.52	9.02
1939	1.64	0.02	0.03	0.00	0.63	1.21	6.46	9.99
1940	1.54	0.03	0.03	0.00	0.54	1.18	6.51	9.83
1941	1.56	0.03	0.05	0.00	0.45	1.15	6.39	9.63
1942	1.57	0.02	0.05	0.00	0.48	1.65	6.72	10.49
1943	1.61	0.02	0.05	0.00	0.48	1.51	6.73	10.40
1944	1.66	0.02	0.06	0.00	0.55	1.86	7.35	11.51
1945	1.80	0.07	0.06	0.00	0.61	1.86	7.77	12.16
1946	1.93	0.07	0.05	0.00	0.64	2.22	8.24	13.15
1947	2.05	0.10	0.06	0.00	0.65	2.91	8.23	14.01
1948	2.10	0.11	0.04	0.00	0.65	2.92	8.52	14.34
1949	2.43	0.14	0.05	0.02	0.62	2.74	9.12	15.13
1950	2.85	0.19	0.06	0.03	0.58	2.56	9.15	15.42
1951	3.03	0.21	0.06	0.04	0.50	1.69	8.56	14.09
1952	3.06	0.19	0.06	0.08	0.50	1.38	8.35	13.61
1953	3.46	0.25	0.05	0.14	0.56	1.24	9.10	14.80
1954	3.73	0.19	0.06	0.19	0.55	1.58	9.53	15.82
1955	3.60	0.12	0.06	0.21	0.63	1.68	9.97	16.27
1956	3.41	0.13	0.06	0.26	0.67	1.92	10.26	16.72
1957	3.41	0.15	0.06	0.32	0.67	2.06	10.26	16.92
1958	3.41	0.14	0.06	0.37	0.72	2.81	10.29	17.81
1959	3.38	0.18	0.07	0.45	0.75	2.56	10.11	17.49
1960	3.41	0.23	0.09	0.49	0.81	2.55	9.81	17.38
1961	3.25	0.20	0.09	0.42	0.89	2.75	8.52	16.12
1962	3.25	0.20	0.10	0.42	0.87	2.59	8.52	15.97
1963	3.27	0.31	0.09	0.42	0.88	2.39	8.53	15.88
1964	3.47	0.52	0.09	0.47	0.90	2.72	8.69	16.86
1965	3.73	0.61	0.11	0.56	0.90	3.00	9.44	18.35
1966	3.98	0.98	0.13	0.62	0.92	3.02	9.37	19.02
1967	4.21	1.29	0.15	0.69	0.93	3.10	9.45	19.81
1968	4.35	1.27	0.18	0.72	0.94	2.99	10.01	20.47
1969	4.39	0.60	0.17	0.82	0.96	2.67	9.28	18.89
1970	4.36	0.30	0.18	0.88	0.94	2.76	9.54	18.97
1971	4.41	0.48	0.32	1.01	0.96	3.61	10.42	21.21
1972	4.48	0.63	0.45	1.05	0.95	3.55	10.72	21.83
1973	4.60	1.04	0.57	1.08	0.94	4.10	10.79	23.11
1974	4.68	1.50	0.70	1.19	0.93	4.02	10.10	23.11
1975	4.80	1.67	0.79	1.29	0.92	4.15	11.50	25.12
1976	4.96	1.65	0.95	1.44	0.92	3.92	11.92	25.75
1977	5.04	1.79	1.24	1.58	0.93	4.68	11.73	27.00
1978	5.18	2.49	1.51	1 80	0.95	4.49	11.29	27.71
1979	5.39	3.01	1.74	1.86	0.94	4.11	10.31	27.36
1980	5.63	3.53	1.63	2.01	0.96	4.33	9.16	27.26
1981	5.81	4.38	1.44	2.08	0.97	4.10	8.53	27.31
1982	5.92	5.00	2.19	2.41	1.00	4.34	7.42	28.27
1983	5.93	5.92	1.61	2.43	1.02	4.25	10.74	31.90
1984	5.91	7.25	1.73	2.75	1.03	4.68	8.05	31.38
1985	5.80	7.75	1.35	2.67	1.04	4.70	8.21	31.52
1986	5.61	8.23	2.10	2.77	1.05	4.72	8.95	33.43
1987	5.45	7.16	1.20	2.63	1.06	5.07	8.96	31.53
1988	5.31	7.08	2.24	2.74	1.03	5.58	10.12	34.09
1989	5.19	6.44	1.68	2.87	1.04	5.78	11.72	34.71
1990	5.07	6.83	1.66	2.78	1.06	5.41	11.57	34.38
1991	4.93	6.14	1.65	2.77	1.06	4.94	10.71	32.19

increased by almost four fold, matched by machinery energy, while fuel energy doubled. Fertiliser energy increased remarkably from 0.03 to over 6 billion kcal, and commercial seed and feed energy increased some fivefold. Pesticides rose from 0.02 to between 1.5 and 2.25 billion kcal, electricity rose from virtually zero to about 2.75 billion kcal, and buildings increased only slightly. The relative importance of each component of input energy is shown in Table 3. Consistently over the

Table 3. Percentage Distribution of Components of Farm Input Energy, 1936 to 1991.

Year	Machinery	Fertiliser	Pesticides	Electricity	Buildings	Seed/Feed	Fuel	Total
1936	18.6	0.3	0.3	0.0	8.8	11.2	60.8	100
1937	18.6	0.3	0.4	0.0	8.4	15.1	57.4	100
1938	17.6	0.3	0.3	0.0	7.5	13.1	61.2	100
1939	16.4	0.2	0.3	0.0	6.3	12.1	64.7	100
1940	15.7	0.3	0.4	0.0	5.5	12.0	66.2	100
1941	16.2	0.3	0.5	0.0	4.7	12.0	66.4	100
1942	14.9	0.2	0.5	0.0	4.5	15.8	64.1	100
1943	15.5	0.2	0.5	0.0	4.6	14.5	64.7	100
1944	14.4	0.2	0.5	0.0	4.8	16.2	63.9	100
1945	14.8	0.5	0.5	0.0	5.0	15.3	63.9	100
1946	14.7	0.5	0.4	0.0	4.9	16.9	62.7	100
1947	14.7	0.7	0.4	0.0	4.7	20.8	58.8	100
1948	14.7	0.7	0.3	0.0	4.5	20.4	59.4	100
1949	16.1	1.0	0.3	0.1	4.1	18.1	60.3	100
1950	18.5	1.2	0.4	0.2	3.8	16.6	59.3	100
1951	21.5	1.5	0.4	0.3	3.6	12.0	60.7	100
1952	22.5	1.4	0.4	0.6	3.6	10.1	61.4	100
1953	23.4	1.7	0.4	1.0	3.7	8.4	61.5	100
1954	23.6	1.2	0.4	1.2	3.5	10.0	60.2	100
1955	22.1	0.7	0.4	1.3	3.9	10.3	61.3	100
1956	20.4	0.8	0.4	1.6	4.0	11.5	61.4	100
1957	20.1	0.9	0.4	1.9	4.0	12.2	60.6	100
1958	19.1	0.8	0.3	2.1	4.0	15.8	57.8	100
1959	19.3	1.0	0.4	2.6	4.3	14.6	57.8	100
1960	19.6	1.3	0.5	2.8	4.6	14.7	56.5	100
1961	20.2	1.3	0.6	2.6	5.5	17.1	52.9	100
1962	20.4	1.3	0.6	2.6	5.5	16.2	53.4	100
1963	20.6	2.0	0.5	2.6	5.5	15.0	53.7	100
1964	20.6	3.1	0.5	2.8	5.3	16.1	51.5	100
1965	20.3	3.3	0.6	3.0	4.9	16.3	51.4	100
1966	20.9	5.2	0.7	3.2	4.8	15.9	49.3	100
1967	21.2	6.5	0.8	3.5	4.7	15.6	47.7	100
1968	21.3	6.2	0.9	3.5	4.6	14.6	48.9	100
1969	23.2	3.2	0.9	4.3	5.1	14.2	49.1	100
1970	23.0	1.6	1.0	4.6	5.0	14.6	50.3	100
1971	20.8	2.3	1.5	4.7	4.5	17.0	49.1	100
1972	20.5	2.9	2.1	4.8	4.4	16.3	49.1	100
1973	19.9	4.5	2.5	4.7	4.1	17.7	46.7	100
1974	20.2	6.5	3.0	5.2	4.0	17.4	43.7	100
1975	19.1	6.6	3.1	5.1	3.7	16.5	45.8	100
1976	19.3	6.4	3.7	5.6	3.6	15.2	46.3	100
1977	18.7	6.6	4.6	5.9	3.5	17.3	43.5	100
1978	18.7	9.0	5.5	6.5	3.4	16.2	40.7	100
1979	19.7	11.0	6.3	6.8	3.4	15.0	37.7	100
1980	20.7	13.0	6.0	7.4	3.5	15.9	33.6	100
1981	21.3	16.0	5.3	7.6	3.6	15.0	31.2	100
1982	20.9	17.7	7.8	8.5	3.5	15.3	26.2	100
1983	18.6	18.6	5.0	7.6	3.2	13.3	33.7	100
1984	18.8	23.1	5.5	8.8	3.3	14.9	25.6	100
1985	18.4	24.6	4.3	8.5	3.3	14.9	26.1	100
1986	16.8	24.6	6.3	8.3	3.1	14.1	26.8	100
1987	17.3	22.7	3.8	8.3	3.4	16.1	28.4	100
1988	15.6	20.8	6.6	8.0	3.0	16.4	29.7	100
1989	14.9	18.6	4.8	8.3	3.0	16.7	33.8	100
1990	14.7	19.9	4.8	8.1	3.1	15.7	33.6	100
1991	15.3	19.1	5.1	8.6	3.3	15.3	33.3	100

years, the greatest contributor to total input energy has been fuel. However, fuel energy declined in relative importance beginning in the late 1950s from highs of about two-thirds of total input energy in the early 1940s to between one-quarter to one-third of total input energy in the late 1980s and early 1990s. The difference was made up by fertiliser (from less than 1% to about one-fifth of total input energy), pesticides (increasing from 1 to 5% of the total), and electricity (from almost none

Table 4. Components of Farm Output Energy, 1936–1991 (10^{12} kcal).

Year	Crop Energy	Livestock	Total
1936			15.39
1937			5.71
1938			18.90
1939			32.04
1940			30.26
1941			20.17
1942			45.35
1943			25.30
1944			35.96
1945			23.69
1946			25.41
1947			23.46
1948			24.15
1949			23.46
1950			30.83
1951	39.88	0.78	40.67
1952	54.33	0.86	55.19
1953	45.81	0.80	46.61
1954	17.32	0.90	18.22
1955	40.62	0.91	41.53
1956	45.81	0.93	46.73
1957	25.05	0.98	26.03
1958	25.33	1.09	26.42
1959	27.70	1.00	28.70
1960	38.92	0.93	39.85
1961	15.20	1.11	16.31
1962	40.28	1.00	41.29
1963	55.43	0.91	56.34
1964	36.67	1.02	37.69
1965	45.87	1.07	46.95
1966	59.81	1.03	60.84
1967	37.72	1.04	38.76
1968	43.35	0.99	44.34
1969	53.86	0.89	54.75
1970	36.93	1.03	37.95
1971	56.41	1.22	57.63
1972	46.55	1.21	47.77
1973	50.46	1.38	51.84
1974	41.39	1.35	42.74
1975	49.59	1.37	50.96
1976	63.94	1.28	65.23
1977	59.95	1.55	61.50
1978	60.94	1.45	62.38
1979	45.67	1.30	46.98
1980	49.41	1.40	50.80
1981	62.81	1.30	64.11
1982	72.03	1.38	73.41
1983	63.60	1.34	64.94
1984	51.44	1.49	52.93
1985	60.32	1.48	61.80
1986	81.39	1.47	82.86
1987	70.50	1.62	72.12
1988	35.82	1.82	37.64
1989	59.46	1.87	61.32
1990	78.51	1.79	80.30
1991	61.31	1.61	62.92

to about 8.5% of total input energy). The relative importance of buildings declined slightly (to about 3% of the total) as one might expect given the reduction in farmsteads, while the relative contribution of machinery rose marginally during the 1950s, 1960s, and 1970s, but ended about where it started, between 15% and 18%, while the relative importance of commercial seed and feed energy increased only marginally.

Farm Output Energy

Table 4 shows the components of farm output energy (in billions of kilocalories). Total output energy reflects the considerable annual variation in crop production. Using broad approximations, total output energy increased at least four-fold over the 55-year period. For the period from the 1950s to the 1990s, crop energy increased by perhaps 150% while livestock energy doubled (see Figure 4), but the relative importance of crops and livestock changed little over this period. Crop energy consistently accounted for 95% to 98% of total output energy.

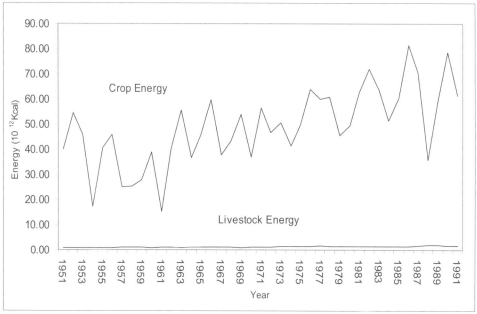

Figure 4. Components of Farm Output Energy, 1951 to 1991.

Trend lines for both total energy outputs and total energy inputs are summarised in Figure 5. The annual variability of output energy is not matched by input energy since some input energy is embedded in relatively fixed capital such as buildings and machinery. Also, farming, by its nature, does not provide the certainty that would allow farmers to accurately scale back other inputs in anticipation of a poor year.

Energy and Labour

The number of farm workers declined dramatically from over 300,000 in the mid-1930s to about 80,000 by the early 1990s. Actually, from the Depression era to the early years of World War II, Saskatchewan farms shed some 100,000 workers. This reduction in farm labour probably reflects the interaction of many factors, such as the exigencies of farming during a major drought and depression, government relief policy during the Depression which sent people back to their home municipalities if they wanted to receive assistance, and the opportunities for alternative employment as a result of the war and internal processes of consolidation in the farm sector.

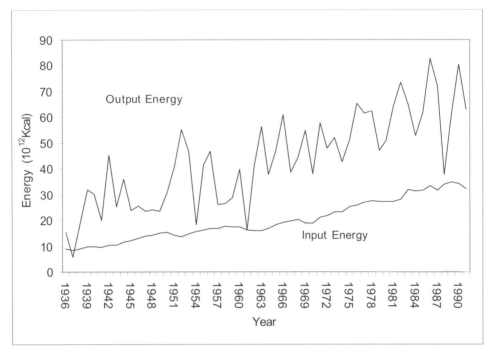

Figure 5. Total Farm Input and Output Energy, 1936 to 1991.

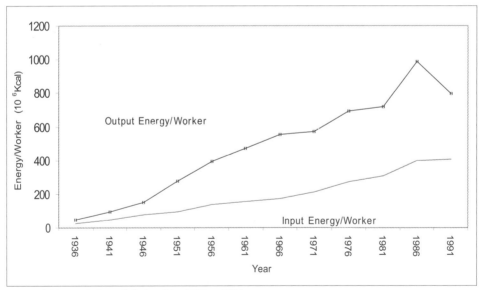

Figure 6. Energy Labour Relationship, 1936 to 1991.

Figure 6 shows both the amount of farm input energy, and of output energy, per farm worker over the study period (here, expressed in millions [10^6] of kilocalories per worker). The output series is characteristically variable but output energy per worker rose between 15- to 20-fold from the mid-1930s to the 1990s. Input energy per worker increased about 15-fold.

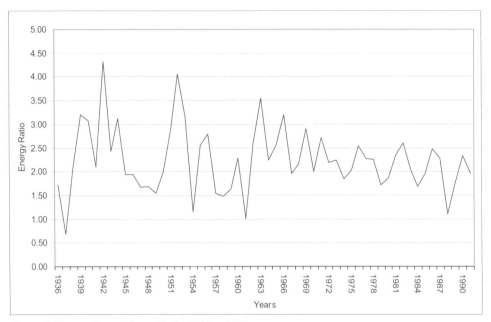

Figure 7. Energy Output/Input Ratio, 1936 to 1991.

Energy Ratio

Figure 7 shows the trend for the ratio of output to input energy (Er) from 1936 to 1991. Recall from Figure 5 that both output and input energy increased. The Er trend also reflects the greater annual variability of output energy. Nevertheless, the Er trend for the 1936–91 period appears to be essentially flat or perhaps slightly downward. For example, by ignoring 1961 and 1988, which were years of abnormally low crop production, one can see a modest downward trend from 1962 to 1987. Even by ignoring the Depression years one finds a modest downward trend from 1939 to 1991.[10]

Conclusions

This study offers tentative answers to some important questions concerning farm energy efficiency, but it also raises other interesting ones, indeed more than can be considered in this article. Also, when contemplating conclusions and possible policy implications, it is important to remember that we are only dealing here with energy efficiency. There is no question that total output from Saskatchewan farms has increased, or that labour productivity has increased. The question is limited to whether or not these increases have been won by using input energy more efficiently. Indeed, it appears that they have not.

Increases in inputs have generated large increases in outputs since the Depression, but it appears that overall, farm technological innovations have not resulted in higher energy efficiency. In contrast, they may have resulted in somewhat lower energy efficiency. The direction of this trend is consistent with the conclusions of Leach (1976) and Pimentel (1980) although, compared to these studies, my findings suggest that the slope of the Er trend over time is more gradually negative or may even be flat. It appears that from the point of view of energy efficiency, for all its technological changes, Saskatchewan farming has been one consistent production system since at least World War II and perhaps earlier. Inputs bearing non-renewable energy have replaced animal power and human labour, and outputs have increased, but energy efficiency has remained constant or declined marginally.

On a historic note, these data do not extend far enough into the past to clearly show the role of horsepower or what is assumed to be the lower input form of frontier farming. Nevertheless, horses were still important during World War II and immediately after (recall, only 39% of farms had a tractor in 1941), so it is interesting to contemplate why the energy ratios for this period are not higher. Perhaps the land base devoted to animal feeding did constrain output energy and hence reduced energy efficiency under this type of farming. Did the higher yields from virgin soil and the lower inputs for settlement farming result in higher energy ratios in World War I and the 1920s, or did the steep learning curve and higher inputs associated with initial capitalisation of farms result in lower energy ratios? The answer awaits another study using a different methodology.

Policy Implications

There are both general and specific policy implications for Saskatchewan farming from this study. The general long term trend has been to produce large volumes of farm outputs and output energy, by using large volumes of non-renewable energy inputs, while reducing the number of farms, as well as the farm population and their communities. This trend is consistent with the structure of prairie farming and its historical function in the Canadian political economy. From the point of view of energy efficiency, it is doubtful that the trend is sustainable. The supplies of non-renewable energy are finite and some are noticeably dwindling (McRae et al., 2000, discuss this and other environmental issues).[11] However, the trend is not inevitable. People, acting through the state, as agribusiness owners, as family farmers, and in other ways, have created the farm structure and influenced the environmental landscape of the prairies. They can also change it.

In the specific and shorter term, one may note that the largest farm energy inputs have come in the form of fuel, machinery, and recently fertilisers. Commercial seed and feed inputs have also increased and may become more important with a rise in GM crops and animals. So, perhaps oddly, if farms increasingly rely on inputs from non-farm firms, then better farm energy efficiency would come from improvements in the energy efficiencies of manufacturing, mining, and even the service sector.

Within the farm industry itself, improved energy efficiency would come through developing, or in some cases redeploying, social organisations, institutions and technologies that would reduce the use of inputs bearing non-renewable energy. This might include the use of machinery cooperatives to reduce the overall need for machines, and consumer cooperatives (see Belcher and Schmutz, this issue). Such co-ops have played an important role in the past although a current collective "amnesia" seems to have left them under-utilised. More generally, farmers may need to reclaim the forgotten wisdom of previous generations of farmers, a knowledge built up from the hands-on experiences of, for example, dry land farming, global markets, and creating "community" (Laidlaw, 2003).

In addition, we need engines with higher fuel efficiencies. If that innovation proves too difficult, at least smaller tractors with better guidance systems could potentially reduce the energy embedded in the tractor and related machinery inventories (Palmer, 2003). Electricity from wind energy would reduce this input component. Better fertilisers and more effective and judicious application methods might reduce the substantial rise in this input. Certainly low input and organic farming will serve to decrease energy inputs (Cushon, 2003).

It is also necessary to question the need for increasing farm outputs.[12] Concerning farm energy efficiency specifically, past increases in outputs have required large outlays of input energy to accomplish. Several programs exist or have been proposed that include limitations on outputs while pursuing other policy goals. "Set aside" programs have been used to protect marginal, wildlife, or sensitive lands. The use of farm land for carbon sequestration might not only require farming to lower its energy use, but it could also serve to transfer much needed income from the non-farm to the farm sector.

Given the energy costs of increasing farm outputs, climate change may force the issue of limiting farm outputs onto the policy agenda. Indeed, the spectre of climate change only makes more obvious the need for policies to take account of agriculture's energy balance (see Sauchyn et al., this issue). For example, energy audits should be required at least for new farming ventures. An important debate has emerged over the energy efficiency of ethanol production using farm grains, usually corn in the United States and wheat in western Canada. Shapouri et al. (2002: 2–3) show how the Er estimates from several studies differ considerably from highs over 1.5, to lows around 0.7 (one Agriculture Canada study they review estimated an Er of 1.4). One important difference between the studies is that the Pimentel work (with the low Er estimates) tries to include all of the embedded input energy, that is, the energy embedded in the manufactured farm machinery, in the ethanol refineries' capital stock, and so forth. Other studies leave many of these inputs out, perhaps assuming that they would be "on the shelf" ready to be used so their energy costs need not be assigned to ethanol production. For effective policy development, we surely need to know the full energy costs of ethanol production, especially if the state is prepared to mandate it for transportation uses and to subsidise its production (Sigurdson, 2003). Compared to an Er of 0.7, an Er of 1.4 indicates a much higher set of costs alongside the assumed benefits of ethanol production and use, such as job creation, and the need for "green" fuel, and makes the whole initiative a much harder bargain. In other words, if energy efficiency is to be a serious farm policy consideration then it will be necessary to vastly improve the way we measure and take account of farm energy.

Notes

1. This article reports on a study of farm energy that was carried out over a 20-year span though the Energy Research Unit at the University of Regina. The research was also supported by grants from the Saskatchewan Science Council and the Prairie Farm Rehabilitation Administration. The Sample Survey & Data Bank Unit at the University of Regina was involved in parts of the study. For their contributions and helpful comments, I would like to thank Lawrence Vigrass, Ewen Coxworth, Denise Kouri, Monty Russell, Brian Kybett, and Huang Kun.

2. The term "energy efficiency" or "Energy Ratio" (Er) has been used by Pimentel (1980) and others to represent a ratio of the amount of metobolisable energy produced per unit of non-renewable energy consumed both directly and indirectly in the production process.

3. Given the customary market arrangements for typical prairie farm commodities, farmers usually haul grains to local elevators, livestock to nearby stockyards, and pick-up consumables such as parts, and some building supplies, and machinery. These minor transportation activities are included in the study but the major costs of transportation are not.

4. This is not to deny the fact that policies and institutional arrangements for the Canadian Prairies were freely borrowed from other frontiers and "white settler" societies. For example, the Dominion Lands Act was modelled after policies used in the American frontier, and other parts of the "national policy," such as tariff policy, were borrowed from policies advanced in Kentucky. Some of these, in turn, were probably borrowed from the earlier practices of Lord Simcoe in Canada West who, no doubt, refined policies from other colonial situations (Watkins, 1991).

5. While the crop land could easily be put to other uses, it is less certain what would have happened to the pasture land. Some of it, although marginal land, could have been turned into crops. Still, the beef cattle herd did increase after World War II by at least 1.5 million.

6. The loss of public data in the 1990s made it difficult to extend the inventory forward. However, other modelling techniques have been used to make such estimates. For examples see MacGregor et al. (2000).

7. This study began with the "kilocalorie" as its unit of measure and I have continued the practice here. To convert it to other units, note that:

   1 kilocalorie   = 1 Calorie   = 1000 calories
                      = 4.186 joules
                      = 3.97 Btu

8. The co-efficients estimated the amount of energy for each constant dollar of expenditure. The co-efficients for each input category were as follows:

   Machinery: 11 million kcal.
   Fertiliser: 15 million kcal.
   Pesticide: 4 million kcal.
   Electricity: 40 million kcal.
   Buildings: 30 million kcal.
   Seed/feed: 55 million kcal.
   Fuel: 40 million kcal.

9. When examining the various time series, one should remember that the estimates for 1961 to 1991 are based on the "inventory" technique, while the estimates for 1936 to 1960 rely on projections using cost and volume indexes.

10. Correlating Er with time yields the following correlation coefficients:

    For the period 1936–91, $r = -.14$ $p = .301$
    For the period 1962–87, $r = -.46$ $p = .017$
    For the period 1939–91, $r = -.28$ $p = .044$

11. In addition, the trend is seriously reducing the farm population and undermining its way of life. Of course some might not see that as an important policy goal as long as food continues to be available for non-farm populations (for example, see the positions taken in Comstock, 1987). Nevertheless, the continued loss of farm families has made family farming less sustainable.

12. The world may not need more food from western nations, particularly when its distribution displaces the ability of poor nations to produce their own food and when it does not curtail hunger (Lappé et al., 1977; Warnock, 1987). There is also the possibility that increasing farm staple exports have not particularly benefited prairie farmers (Qualman and Wiebe, 2002). However, these are issues beyond the scope of this article.

References

Comstock, G. (ed.). 1987. *Is There A Moral Obligation to Save the Family Farm?* Ames: Iowa State University Press.

Cushon, I. 2003. "Sustainable Alternatives for Saskatchewan Agriculture: A Farmer's Perspective." Pp. 223–236 in H.P. Diaz, J. Jaffe and R. Stirling (eds.), *Farm Communities at the Crossroads Challenge and Resistance.* Regina: Canadian Plains Research Center.

Fowke, V. 1957. *The National Policy and the Wheat Economy.* Toronto: University of Toronto Press.

Government of Canada. 1969. *Canadian Agriculture in the Seventies: Report of the Federal Task Force on Agriculture.* Ottawa: Information Canada.

Hulse, J. 1995. *Science, Agriculture & Food Security.* Ottawa: NRC Research Press.

Jensen, N.E. 1977. *Total Energy Budgets for Selected Farms in Western Canada.* Olds, AB: Jensen Engineering Ltd.

Lappé, F.M., J. Collins and C. Fowler. 1977. *Food First: Beyond The Myth of Scarcity.* Boston: Houghton-Mifflin.

Laidlaw, S. 2003. *Secret Ingredients The Brave New World of Industrial Farming.* Toronto: McClelland & Stewart Inc.

Leach, G. 1976. *Energy and Food Production.* Guilford, Surrey: IPC Business Press Limited.

MacGregor, R., R. Lindenbach, S. Weseen and A. Lefebvre. 2000. "Energy Use." Pp. 171–177 in T. McRae et al. (eds.). *Environmental Sustainability of Canadian Agriculture Report of the Agri-Environmental Indicator Project.* Ottawa: Agriculture and Agri-Food Canada.

Marchetti, C. 1979. *On Energy and Agriculture: From Hunting-Gathering to Landless Farming.* Laxenburg, Austria: International Institute for Applied Systems Analysis.

McRae, T., C. Smith and L. Gregorich (eds.). 2000. *Environmental Sustainability of Canadian Agriculture Report of the Agri-Environmental Indicator Project.* Ottawa: Agriculture and Agri-Food Canada.

Mathieson, G. 1977. "Present Use of Energy on Canadian Farms: A Reaction." Pp. 46–51 in *Saskatchewan Agriculture. Energy Conservation in Agriculture Proceedings of Seminars.* Regina: Saskatchewan Agriculture.

Moseley, W. and C. Jordan. 2001. "Measuring Agricultural Sustainability: Energy Analysis of Conventional Till and No-Till Maize in the Georgia Piedmont." *Southeastern Geographer* 41, no. 1: 105–116.

Palmer, R. 2003. "Tractor Guidance Systems and the Family Farm." Pp. 259–266 in H.P. Diaz, J. Jaffe and R. Stirling (eds.), *Farm Communities at the Crossroads Challenge and Resistance.* Regina; Canadian Plains Research Center.

Pimentel, D. (ed.). 1980. *Handbook of Energy Utilization in Agriculture.* Baton Raton, FL: CRC Press, Inc.

Pimentel, D., L. Hurd, A. Bellotti, M. Forster, I. Oka, O. Sholes and R. Whitman. 1973. "Food Production and the Energy Crisis." *Science* 182: 443–449.

Pimentel, D. and M. Burgess. 1980. "Energy Inputs in Corn Production." Pp. 67–84 in D. Pimentel (ed.), *Handbook of Energy Utilization in Agriculture.* Baton Raton, FL: CRC Press, Inc.

Qualman, D. and N. Wiebe. 2002. *The Structural Adjustment of Canadian Agriculture.* Ottawa: Canadian Centre for Policy Alternatives.

Rowe, Stan. 1990. *Home Place Essays on Ecology.* Edmonton: NeWest Publishers Limited.

Saskatchewan Agriculture and Food. 1979–2001. *Agricultural Statistics.* Regina: Policy Branch, Saskatchewan Agriculture and Food.

Saskatchewan Royal Commission on Agriculture and Rural Life. 1955a. *Report No. 2: Mechanization and Farm Costs.* Regina: Queen's Printer.

——. 1955b. *Report No. 8: Farm Electrification.* Regina: Queen's Printer.

——. 1957. *Report No. 14: A Program of Improvement for Saskatchewan Agriculture and Rural Life.* Regina: Queen's Printer.

Slesser, M. 1986. "Energy Balance in Agriculture: The Developed World." *Natural Resources and the Environment* 20: 47–56.

Shapouri, H., J. Duffield and M. Wang. 2002. *The Energy Balance of Corn Ethanol: An Update. Agricultural Economic Report No. 813.* Washington: U.S. Department of Agriculture, Office of the Chief Economist, Office of Energy Policy and New Uses.

Shepard, R.B. 1994. "American Influence on the Settlement and Development of the Canadian Plains." PhD dissertatio, University of Regina.

Sigurdson, K. 2003. "Scrap Plan to Subsidize Ethanol Production." *The Union Farmer Quarterly* 9, no. 2: 20–21.

Southwell, P. and T. Rothwell. 1977. *Analysis of Output/Input Energy Ratios of Food Production in Ontario.* Guelph, ON: School of Engineering, University of Guelph.

Stirling, B. 1979. *Use of Non-Renewable Energy on Saskatchewan Farms: A Preliminary Study.* Regina: Saskatchewan Science Council.

Stirling, B. and Huang Kun. 1992. *An Energy Inventory for Saskatchewan Agriculture 1976–1990 Final Report.* Regina: Energy Research Unit, University of Regina.

Timbers, G. 1977. "Present Use of Energy on Canadian Farms." Pp. 24–45 in *Energy Conservation in Agriculture Proceedings of Seminars.* Regina: Saskatchewan Agriculture.

Warnock, J. 1987. *The Politics of Hunger: The Global Food System.* Toronto: Methuen.

Watkins, M. 1991. "The 'American System' and Canada's National Policy." Pp. 148–157 in G. Laxer (ed.), *Perspectives on Canadian Economic Development Class, Staples, Gender and Elites.* Toronto: Oxford University Press.

White, C.O. 1976. *Power for a Province: A History of Saskatchewan Power.* Regina: Canadian Plains Research Center.

CHAPTER 4

The Changing Prairie Social Landscape of Saskatchewan: The Social Capital of Rural Communities

Polo Diaz and Mark Nelson

Introduction

Rural prairie communities have faced significant tribulations during the last few decades. As a result, many of them have disappeared while many others have witnessed a deterioration of community life and face an uncertain future. One of the most pressing problems faced by rural prairie communities has been the centralization of goods and services and associated infrastructure, which has reduced the infrastructure and number of community outlets in rural areas and increased the concentration of these outlets in a few urban localities. As Mitchell (1975) points out, the process started in the 1960s, when private companies began to centralize their services in large, urban localities and to close the doors of their outlets in the rural communities. In the next two decades, public services such as postal offices, hospitals, and schools followed the same path as many companies (e.g. the old grain elevators) disappeared from the rural landscape. At the same time, rural communities have been subjected to a continuous process of rural depopulation. According to Statistics Canada census data, the total rural population of Saskatchewan in 1981 was 405,147. By 2001, that number had steadily decreased to 349,897 (Statistics Canada, 2001). This continuous decrease is a problem for the viability of the rural communities in the long term. The exodus of many families, especially of a predominantly young and single population, adversely affects rural communities, reducing their population to a point at which many of them cannot be sustained.

Over the last century, farming has been the key to the prairie rural economy and to the rural community. Thus, the farm crisis of the last few decades has also threatened the viability of the rural community. Financial hardships and farm bankruptcies have become part of the everyday life of farmers and their communities. Net farm income has declined in recent decades, preventing many rural families from reaching a minimum standard of living. For example, in 1995, net farm income represented 35.5% of the total household income for farm households in Saskatchewan. By 2000, this percentage had decreased to 27.2% (Statistics Canada, 2001). A side effect of upheavals in the farm economy has been the differentiation of the traditional family farm (Conway and Stirling, 1985), fostering the creation of more large, heavily capitalized farms and more small farms. As Harder (2001) argues, this process of differentiation has resulted in increased competition and a greater diversity of interests among farmers.

The process of rural restructuring has been intensified by globalization and its related policies, which view rural areas as hinterland appendages to urban centres. Several authors (Lawrence et al., 2001; Epp and Whitson, 2001) point out that neo-liberalism has brought further economic decline to the rural sector, an intensification of consumer culture, environmental degradation, and a more pronounced rural-urban division of labour. All of these changes have impacted the daily life of rural community residents by reshaping customs, norms, and local institutions.

Do these dramatic changes in the economic and social landscape of rural communities reflect a loss of social cohesion within rural communities? Has the "community spirit" disappeared from rural life? How has the social fabric of communities been affected by these economic and social upheavals? This article focuses on the patterns of social capital in a group of rural communities in southern Saskatchewan by examining patterns of trust and participation in informal and formal organizations and networks within these communities. The discussion of these patterns is based on survey data collected in 1987 and 2002. These surveys employed instruments that incorporated many of the same questions, allowing for a comparison of the two datasets. However, as the two surveys were conducted in different communities (only the town of Naicam is common to both) as well as in different decades, this comparison must be made carefully. We are aware of the difficulty of making any definitive conclusions on the evolution of social capital based on differences in these datasets. Our purpose is more focussed on the search for general trends rather than demonstrating the direction of a historical process. Because the 1987 and 2002 project teams both attempted to select communities that reflect the diversity of rural Saskatchewan, a comparison of these datasets is adequate for that purpose.

What is Social Capital?

During the last decade, social capital has become a prominent concept in both the academic and policy-making communities. The concept refers to all features of social life (networks, social trust, norms of reciprocity, and participation in organizations) that promote the coordination of collective actions (Putnam, 2001). These features promote social networks that facilitate both the coordination and development of social cooperation around tasks that require collective efforts and social integration. It is in these terms that social capital has been defined as a fundamental component of the social cohesion process (Reimer, 2001). The development of the concept is found mainly in the works of sociologists like Bourdieu and Coleman (for a discussion of the historical development, see Schuller et al., 2000). In the early 1990s, the concept became part of the political lexicon as a result of the work of Robert Putnam, who argued that the economic and institutional performance of a country was dependent on the level of social capital, an argument that had a strong influence on the agendas of many national and international bodies (Putnam, 1993, 1996, 2000, and 2001). Since then, the concept has become an important component in the agendas of many public agencies, such as the World Bank.

There are different understandings of what constitutes social capital depending on different theoretical perspectives (for a detailed discussion of the definition of social capital, see Office for National Statistics, 2001). In a Neo-Marxist perspective, represented by Bourdieu, social capital is understood in the context of relationships of power and resource distribution. In a liberal perspective, exemplified in the work of Putnam (2001), social capital is just a function that facilitates individuals' action in a large number of structural entities. Different perspectives also emphasize different dimensions of social capital. Putnam, for example, talks about "bonding," "bridging," and "linking" forms of social capital. Bonding social capital makes a community internally cohesive whereas bridging social capital strengthens the links that bring different communities and different institutional levels together. It is logical that these two forms of social capital should be used *in tandem* in any strategy dealing with the viability of communities. The problem is that we do not have a theoretical agreement or an empirical understanding about how bridging capital works and how it is affected by bonding social capital. We can easily think of situations, for example, in which bonding capital becomes stronger in a community because of a lack of bridging capital. Similarly, not all forms of social capital contribute to better integration and social cohesion. There is a consensus, however, that the definition of social capital needs to emphasize the role of social networks and civic norms.

What makes social capital attractive to policy-makers and stakeholders is its capacity to act as a bridge between the micro level of individuals and the macro level structures of society (Lechner, 2002). Social capital is a phenomenon that takes place at the level of community, with consequences that impact other processes, such as the distribution of political and economic power. Moreover, social capital seems to be strongly related to human capital, a concept widely used in policy thinking, especially in the areas of education and training (Field et al., 2000; Cote, 2001; Schuller, 2001). While human capital is applied to the individual, and refers to the knowledge, skills, and competences embodied in each person, social capital deals with formal and informal relationships among individuals and the networks that emerge from these relationships.

The social nature of social capital is expressed in forms like family structures, informal networks, formal institutions, and neighbourhood organizations. In these terms, social capital could be defined following Hirschman (1984), as "social energy," the force that keeps a community organized and cohesive, or as a social asset attained through participation in a social group. Social capital is accumulated within the community, usually as an unintended and even unanticipated consequence, through social relationships and interactions (Maskell, 2000). In these terms, social capital refers to resources embedded in networks, rather than to characteristics of individuals.

A large number of empirical works indicate that social capital could have positive implications for a variety of social and economic issues. Social capital correlates with strong economic performance (Woolcock, 2001), lower crime (Putnam, 2001), better health (Veenstra, 2001), immigrant adaptation (Lauglo, 2000), tolerance (Putnam, 2001), and community development (MacGillivray and Perry, 2000), among other important social and economic dimensions. Social capital, and the collective activity that it fosters, can support survival strategies for small farms, such as the establishment of sustainable agricultural practices, and thereby improve the economic situation of the entire community (Flora, 1995).

Social capital also contributes to the welfare of individuals by facilitating the dissemination of knowledge about the availability of jobs and by providing an information channel for markets, products, and economic opportunities. Knowing other people, socializing with them, and doing things together can facilitate an understanding of the rules of labour markets and of the multitude of instruments that could be used for securing a decent family income. It also constitutes an important tool for social support, facilitating the provision of health care, child-care, and transportation (Reimer, 1997).

The social capital concept requires careful consideration since it is not a panacea for rural health. Its increasing popularity and widespread use have created a situation where its original meaning and heuristic value are being seriously tested (Portes, 1998). On a theoretical level, the concept requires further discussion and analysis. Most of the work in this area has been focused on the effects of social capital. Less attention has been given to the mechanisms that allow for the existence of this capital in its many forms (Glaeser, 2001), thereby limiting the development of policy actions that could contribute to increasing the levels of social capital. While conceptual refinement is required, social capital has the potential to be a fundamental tool for dealing with the adversities and opportunities created by the social and economic changes of recent modernity.

With these issues in mind, our examination of the data, rather than concentrating on the effects of social capital, was geared toward gaining a basic understanding of the constituent elements of social capital in our subject communities. Our analysis focuses on the two central dimensions that characterize social capital: (1) the existence of social norms that promote cooperation and collaboration among individuals and (2) socially-oriented behaviour.

Methods

Our discussion is based on analysis of two survey data sets. The first survey, which took place in 1987, was part of a study of families living in four rural communities in southern Saskatchewan: Coderre, Stewart Valley, Naicam, and Wishart. The study was part of a larger one that focused on a broad range of topics related to farm, household, and community relations. The survey involved two stages: (1) phone interviews with 304 respondents, and (2) a mail-out questionnaire to those same respondents. The sample was limited to families engaged in farming, so the results do not take into consideration the situation of non-farm families living in the community.

The second survey, carried out in 2002, is part of a larger study about the social cohesion of Saskatchewan rural communities. Approximately 500 individuals living in and around six rural communities (Balcarres, Carlyle, Craik, Eastend, Naicam, and Willow Bunch) were surveyed. The survey was designed to explore the issues of participation in organizations, services used in the community, levels of trust, attitudes about the level of social cohesion of communities, income, labour force, and educational activities. The second survey was directly focused on issues of social cohesion, offering a more comprehensive picture of the forces of integration and exclusion that characterize the rural community. It gathered information from both farm and non-farm households living in the community or its immediate vicinity.

Results and Discussion

As mentioned above, we viewed social capital as a phenomenon characterized by two dimensions: (1) the existence of socially-oriented norms, such as solidarity, reciprocity, and trust, and (2) socially-oriented behaviour, including participation in informal organizations and networks.

Socially-Oriented Norms

As the basis for mutual support and help, trust is considered to be a fundamental component of social capital. Communities with a rich stock of trust and social networks are in a stronger position to deal with crises, tensions, and challenges. Trust fosters reciprocity and the existence of networks. It also strengthens social relations and promotes shared values and feelings of common identity and a sense of community belonging.

The 2002 survey asked two basic questions regarding trust. The first asked respondents if they would agree with the idea that most people outside the community ("outsiders") could be trusted or if they felt a need to be careful in dealing with them. The second question was similar, but oriented to measure the level of trust with regard to people living in the community, the "insiders." Responses were strongly positive for insiders, with 87% of the respondents stating that members of the community could be trusted. The level of trust for outsiders was lower, but still significant. Almost 60% of the respondents expressed their trust of people from outside the community. This does not mean that trust is a homogenous phenomenon in these communities. Our data indicate that the degree of trust varies between different communities, age groups, educational levels, economic activities, and genders. Men more than women, for example, seem to be predisposed to trust insiders (91% of the men versus 84% of women) and outsiders (64% versus 57%). Overall, however, members of the community seem to be aware of the relevance of trust for the cohesion of the communities. Respondents were asked about the importance of several issues that may affect the ability of the community "to stick together" and get things done, including the existence of trust. Eight out of ten respondents agreed that trust is important for the community's cohesion.

Others variables in the 2002 survey show similar trends regarding peoples' perceptions of social relations within the community. For example, respondents were asked to express their level of agreement regarding several statements about how people relate to each other in the community. Over 80% of respondents were positive to all of the statements expressing strong relationship between the respondent and the community (Table 1). These results show the prevalence of norms that

Table 1. Perceptions of Social Relations Within the Community.

Statements	Strongly Agree	Agree
If I feel like talking, I can find someone in the community to talk with.	43.5%	50.7%
If there were a serious problems in this community, people here would get together and solve it.	35.9%	52.4%
My friends in this community are part of my everyday activities.	32.2%	48.2%
I feel a real sense of belonging here.	42.0%	50.9%
People in this community share many of the same values.	31.7%	58.9%
If I am upset about something personal, there is no one in this area that I can turn to.	1.7%	7.3%

Table 2. People Respondents Would be Uncomfortable Having as Neighbours.

Groups	Mentioned
People of different race	3.3 %
Native people	13.0 %
Jews	2.9 %
Recent immigrants	7.3 %
Homosexuals	32.4 %
People with AIDS	30.0 %
Drug addicts	72.0 %
Heavy drinkers	56.6 %
Emotionally unstable people	50.4 %

promote the existence of a strongly integrated and internally cohesive set of communities.

The 2002 survey also measured the degree of social exclusion with regard to several groups of people. These groups included ethnic minorities (such as people of different race, Native people, Jews, and recent immigrants) and social minorities (homosexuals, people with AIDS, drug addicts, heavy drinkers, and emotionally unstable people). Respondents were asked to select those groups they felt uncomfortable having as neighbours. The results indicate that social rejection is focused mostly in the social minority categories, with at least one-third of the respondents rejecting them, and in the specific case of drug addicts, more than two-thirds (Table 2).

High levels of trust, value sharing, and a strong sense of belonging are fundamental characteristics of these communities, but these norms of socially oriented behaviour are accompanied by processes of social exclusion. The social exclusion patterns, focussing on social minorities instead of ethnic minorities, reflect the traditional social conservatism of rural communities. As such, the "traditional" sorts of limited social exclusion, along with the high levels of trust and shared values, indicate to us that these communities have a capacity to develop effective levels of social capital.

Participation in Formal and Informal Organizations

If socially oriented norms, such as trust, represent the subjective dimension of social capital, then participation in organizations is the concrete, objective form that social capital assumes in a community. Given that social capital represents the extent to which people work and act together, participation in formal and informal organizations or networks is always considered to be a fundamental dimension of the concept. This dimension involves solid and permanent relationships with relatives, friends, and co-workers, as well as active participation in community organizations.

The crucial feature of participation in terms of social capital is connectedness. Organizations, formal and informal, connect people with each other, allowing for a more cooperative and generous orientation to action, facilitating the establishment of collective objectives and forms of mutual support.

Formal and Informal Networks

As noted above, participation in formal and informal networks is an important indicator of social capital. Our analysis of this form of social capital concentrates on two issues: (1) contributions from the community to the respondents' household and (2) the engagement of the respondents with the community. To that end, we examined a number of relevant, directly comparable questions from the two surveys and we noted some interesting differences, especially in the network participation of women. Again, we must keep in mind that we are dealing with data collected from different communities as well as different decades, so the results discussed below represent an exploration of broad trends in the operation of the mechanisms of social capital in rural Saskatchewan, rather than any specific historical processes.

Time Spent with Friends: Time spent with relatives and friends played an important part in the network of informal connections existing in the communities in both time frames. Looking at how often respondents spent time with these friends, we find that in the 1987 survey 73.7% of the respondents said that they spent time with friends weekly, while in 2002 that number was significantly higher (83.4%).

In 2002 there was no significant relationship between gender and the amount of time spent with friends, but that was not the case in 1987. Data from 1987 indicate that about 70% of men spent time with friends on a weekly basis, compared to women at 80%. In 2002 men nearly equalled women in this category at over 80%. This result could indicate that men came to value informal network associations as much as women do, or it may be that men were able to spend more time with friends recently because of increased retirement in an aging community.

Church Activity: Church attendance and/or participation in church-based activities were significantly lower in the 2002 survey. In 2002, the number of people reporting weekly or monthly church attendance/participation was 10% lower than in 1987, while those attending/participating once or twice a year or not at all was 10% higher.

The relationship between church activity and gender was not statistically significant in 1987, but was in 2002. Participation was lower overall in the 2002 data than in the 1987 data, but the difference in the participation of women is only slight compared to that of men. About 5% fewer women attended church or church-based events weekly in the more recent data, approximately 1% fewer attended monthly, and only 5% more reported yearly participation or no participation. By contrast, the 2002 data indicate that 8% fewer men were attending church or church-based events on a weekly basis and 10% fewer were attending monthly compared to 1987. The number of men who say that they attended/participated in church activities only a few times a year or not at all was 37.3% in the 1987 data and 54.5% in 2002.

Sports, Volunteerism, and Service Organizations: Levels of overall participation in sport clubs, voluntary activities, and service organizations in the communities were higher in 2002 than in 1987, but the amount of time that the 2002 respondents were able to spend on this participation appears to have been limited. The number of respondents who reported no involvement or only limited involvement in such organizations was lower in 2002 (39% in 1987 versus 29% in 2002), but the number of people reporting weekly involvement was nearly the same in both surveys at just over 40%. The most dramatic difference is in the number of respondents reporting monthly involvement in sport, volunteer, or service organizations. The percentage of 2002 respondents who said that they spent time at these sorts of organizations once or twice a month was nearly double than reported in 1987 (16.4% in 1987 versus 29.8% in 2002).

Looking at the changing levels of sport, volunteer and service organization involvement broken down by gender, we find a different pattern for men and women. For men, the difference in the monthly involvement category comes at the expense of the "yearly or not at all" categories. So for men the differing pattern really does indicate a higher level of participation in 2002 compared to 1987. For women, however, the difference in the monthly category reflects corresponding differences in the weekly category. Weekly participation by women was 50% in 1987 and 40% in 2002, while monthly participation was 15% and 31% respectively. The number of women who indicated only a yearly involvement, or no involvement, only differed by about 5% between the two datasets, while the percentage of men reporting yearly or no participation was 13% lower in 2002 than in 1987. For women, overall participation in these sorts of organizations was at similar levels in both datasets, but they were devoting less time to these organizations in 2002.

Number of Organizations: The 2002 survey respondents belonged to, or participated in, fewer organizations than their counterparts in the 1987 survey. The greatest difference is in the participation of women. In 1987, 60% of women belonged to four or more organizations compared to 50% for men. In fact, fully half (50%) of the women reported belonging to five or more organizations, while only 25% of men were involved in five or more. In 1987, the relationship between sex and number of organizational involvements was statistically significant, but in 2002 it was not. Of the female respondents in the 2002 survey, 29% reported belonging to four or more organizations, compared with 22% of the males. The percentage of males and females belonging to five or more organizations was even more similar at 15% and 17% respectively.

If the gender differences we found between 1987 and 2002 are indicative of a larger trend, then our analysis suggests that the network participation and organizational involvement of men and women in rural communities have become more alike than they were in the past.

Networks of Solidarity

If differences between these two datasets reflect broad trends in rural Saskatchewan, then the operation of mutual support networks appears to have undergone a transformation between 1987 and 2002. The 1987 survey concentrated its emphasis, with regard to mutual support, on the contributions of various community members to local farms. Data from that survey indicate that only relatives made contributions to farms. No respondent listed non-relatives as contributing to the farm, suggesting that mutual support networks were based exclusively on kinship rather than friendship. Data from the 2002 survey, by contrast, indicate a much broader network of social solidarity. Because of the broader focus of the 2002 survey, it contained questions related to both farm and non-farm labour exchanges. These questions range in content (friendship and advise, gifts of money, etc.). For each item, the respondent was asked to indicate if she or he received that type of help, provided that type of help, or both. In a marked contrast to the 1987 data, analysis of these labour exchange variables shows a network of mutual support encompassing both relatives and non-relatives. Indeed, more people reported receiving help from non-relatives (86.3%) than from relatives (74.7%), and more people reported providing help to non-relatives (92.1%) than to relatives (84.3%). The same trend prevails for farm households in the 2002 data.

In examining farm-related labour exchanges among farming households, we again find a network of solidarity that is not exclusively kinship-based.

Diaz and Gingrich (1992) note a definite trend in the 1987 data towards contributions within kinship-based networks of mutual support. Contributions tended to flow toward younger farm families that were trying to establish themselves. This trend does not appear to persist in 2002. The relationship between age and receiving farm help from both relatives and non-relatives is statistically significant, but that help does not seem to flow from younger people to older people. Rather, it seems to flow between people of the same broad age cohort. In terms of help received from both relatives and non-relatives, 44% of recipients were between 25 and 44 years of age and another 30% were between 45 and 54. The percentages are nearly identical for those respondents providing help.

Do these differences in the reported patterns of networks of solidarity illustrate actual changes in resource-sharing behaviours of rural residents? Do we have a qualitative and quantitative difference between the ways in which solidarity functioned in 1987 and 2002? While many of these differences may not be related to qualitative differences but rather to artefacts produced by the 1987 survey, they still reflect the presence of strong networks of solidarity that would have contributed enormously to the viability of families and communities.

Conclusions

The data outlined here show that in spite of the serious problems of rural Saskatchewan, social capital is an integral part of rural communities. The strong levels of networking and participation, combined with high levels of trust and internal integration, indicate that the main elements of social capital are present in these communities in abundance. With high levels of social capital, the people of these communities should have the capacity to respond and adapt to the many problems that face rural Saskatchewan at present.

These results, however, should not be considered definitive for four reasons. First, since the trends we discuss are the product of surveys that were oriented to obtain general information on issues relevant to the community, the findings are tentative and preliminary. Also, given that the trends are based on a comparison of data collected in two different sets of communities as well as in two different time frames, our discussion is at the level of broad operation trends of social capital mechanisms in rural communities. Thus, it does not have the precision needed to outline specific historical processes within specific communties.

Second, our examination of social capital in these communities has emphasized the main constituent elements of social capital, but the analysis does not take into consideration the effects of social capital on the social fabric of the communities. Social capital could exist under a variety of social and economic conditions without necessarily having a positive influence on these conditions.

Third, our discussion has focussed on bonding capital and not linking capital. While our subject communities seem to possess a high level of social capital, we must bear in mind that that social capital is operating in a context of rural depopulation, an aging population, and an agricultural economy characterized by centralization of goods and services. In this context, the bonding form of social capital may be operating strongly but the bridging form may not, making it difficult for a community to ally itself to neighbouring communities and importantly, to access outside capital (Flora, 1995). These contextual limits on the potential of social capital to address community problems may have serious implications for the future survival of these communities.

Finally, a distinction must be made between levels of social capital and the quality of social capital. For example, human capital is a component of social capital, and this level must be taken into consideration. In other words, the whole is more than the sum of its parts where social capital is concerned, but the value of the whole is still limited by the resources of its constituents. When we think of social capital as a means of addressing community problems, we must consider not only the potential for collective action in a particular community but also who the people involving themselves are and what everyone is bringing to the table. Flora (1995) tells us that the ability to locate and mobilize resources is fundamental to the creation of the kind of social capital that has the capacity to address the problems facing rural communities. Future investigation, therefore, should concentrate on the link between micro- and macro-level social structures that make social capital so attractive to policy-makers and stakeholders.

Our preliminary investigation of the constituent elements of social capital in our subject communities has yielded more questions than definitive conclusions. Indeed, that was its purpose. Within the larger "Social Cohesion" project, the 2002 survey data were intended to provide a broad quantitative context for other, more focussed research efforts, both quantitative and qualitative. The patterns we present in this paper invite closer exploration by the ethnographic analyses and the more elaborated surveys taking place through the larger project, in order to develop solid explanations. As such, the data collected in 2002 and the differences identified in comparisons with the 1987 survey provide us with a solid contextual foundation. The data show high levels of social capital within the communities examined in 2002 and hint at some intriguing patterns in the evolution of social capital in rural settings.

In spite of the challenges these communities have faced in recent decades, they are still very cohesive localities. Perhaps, in their accumulated social capital, we can find responses to future challenges imposed by the larger processes of the outside world. The key to solid adaptive measures may be strong local organizations that are able to engage the majority of local residents in collective activities that help families and individuals adjust to the uncertainties of new social and economic conditions. The communities we compared would seem to have a solid foundation, in the form of social capital, for building effective responses to the challenges of recent modernity. Through this initial examination of that social capital, we have built a solid foundation for examining and understanding their responses.

Acknowledgements

The authors thank Stan Morse, Grace Shelton, and one anonymous reviewer for helpful comments on earlier drafts. This publication is the product of the research activities of the multidisciplinary project "Rural Adaptation and Social Cohesion for Sustainable Development in the Prairies." The project, housed in the Canadian Plains Research Center of the University of Regina, has been supported by the Social Sciences and Humanities Research Council of Canada.

References

Cote, S. 2001. "La contribution des capacities humaines et sociales," *Isuma* 2, no. 1: 25–33.

Conway, J. and R. Stirling. 1985. "Fractions Among Prairie Farmers." Paper presented to the annual meeting of the Canadian Political Science Association, Montreal.

Diaz, P. and P. Gingrich. 1992. "Crisis and Community in Rural Saskatchewan." In David Hay and Gurcharn Basran (eds.), *Rural Sociology in Canada*. Toronto: Oxford University Press.

Epp, R. and D. Whitson. 2001. "Introduction: Writing off Rural Communities." In R. Epp and D. Whitson (eds.), *Writing Off the Rural West*. Edmonton: The University of Alberta Press/Parkland Institute.

Field, J., T. Schuller and S. Baron. 2000. "Human and Social Capital Revisited." In S. Baron, J. Field and T. Schuller (eds.), *Social Capital. Critical Perspectives*. NY: Oxford University Press.

Flora, C. 1995, "Social Capital and Sustainability," *Research in Rural Sociology and Development* 6: 227–246.

Glaeser, E. 2001. "The Formation of Social Capital," *Isuma* 2, no. 1: 34–40.

Harder, C. 2001, "Overcoming Cultural and Spiritual Obstacles to Rural Revitalization." In R. Epp and D. Whitson (eds.), *Writing Off the Rural West*. Edmonton: The University of Alberta Press/Parkland Institute.

Hirschman, A.,1984. *Getting Ahead Collectively: Grassroots Experiences in Latin America*. New York: Pergamon Press.

Lauglo, J. 2000. "Social Capital Trumping Class and Cultural Capital? Engagement with School Among Immigrant Youth." In S. Baron, J. Field and T. Schuller (eds.), *Social Capital. Critical Perspectives*. NY: Oxford University Press.

Lawrence, G., M. Knuttila and I. Gray. 2001. "Globalization, Neo-Liberalism and Rural Decline in Australia and Canada." In R. Epp and D. Whitson (eds.), *Writing Off the Rural West*. Edmonton: The University of Alberta Press/Parkland Institute.

Lechner, N. 2002. "El capital social como problema cultural," *Revista Mexicana de Sociologia* 64, no. 2: 91–109.

MacGillivray, A. and W. Perry. 2000. "Local Social Capital: Making it Work on the Ground." In S. Baron, J. Field and T. Schuller (eds.), *Social Capital. Critical Perspectives*. NY: Oxford University Press.

Maskell, P. 2000. "Social Capital, Innovation, and Competitiveness." In S. Baron, J. Field and T. Schuller (eds.), *Social Capital. Critical Perspectives*. NY: Oxford University Press.

Mitchell, D. 1975. *The Politics of Food*. Toronto: James Lorimer.

Office for National Statistics (UK). 2001. *Social Capital. A Review of the Literature*. London: Social Analysis and Reporting Division.

Portes, A. 1998. "Social Capital: Its Origins and Applications in Modern Sociology," *Annual Review of Sociology* 24: 1–24.

Putnam, R.D. 1993. *Making Democracy Work*. Princeton: Princeton University Press.

——. 1996, "Who Killed Civic America," *Prospect* (March): 66–72.

——. 2000. *Bowling Alone: The Collapse and Revival of American Community*. New York: Simon Schuster.

——. 2001. "Social Capital: Measurement and Consequences," *Isuma* 2, no. 1: 41–52.

Redclift, M. 1986. "Survival Strategies in Rural Europe. Continuity and Change," *Sociologia Ruralis* 19, no. 3.

Reimer, B. 1997. "Informal Social Networks and Voluntary Associations in Non Metropolitan Canada." In R. Rounds (ed.), *Changing Rural Institutions*. Brandon: Canadian Rural Reestructuring Foundation/Rural Development Institute.

——. 2001. "Understanding and Measuring Social Capital and Social Cohesion," *New Rural Economy* Webpage: ftp://132.205.151.25:21/spss_2001/3csmeasures1.pdf

Schuller, T. 2001. "The Complementary Role of Social and Human Capital," *Isuma* 2, no. 1: 18–24.

Schuller, T., S. Baron and J. Field. 2000, "Social Capital: A Review and a Critique." In S. Baron, J. Field and T. Schuller (eds.), *Social Capital. Critical Perspectives*. NY: Oxford University Press.

Statistics Canada. 2001. "Farm Population, by Provinces (2001 Census of Agriculture)." Last modified: 2004–05–30. Available from: http://www.statcan.ca/english/Pgdb/agrc42i.htm [Accessed July 5, 2004]

——. 2001. "Farm Households by Income, by Provinces (2001 Census of Agriculture)." Last modified: 2004–05–30.Available from: http://www.statcan.ca/english/Pgdb/agrc36i.htm [Accessed July 5, 2004]

Veenstra, G. 2001. "Social Capital and Health," *Isuma* 2, no. 1: 72–81.

Woolcock, M. 2001, "The Place of Social Capital in Understanding Social Social and Economic Outcomes," *Isuma* 2, no. 1: 11–17.

CHAPTER 5

Management of the Prairie Landscape Through Strategic Consumer-Producer Cooperatives

Ken Belcher and Josef K. Schmutz

Introduction

Agricultural producers in the northern Great Plains are experiencing decreasing net incomes with rising input costs and falling commodity prices. As a result, farmers are being forced to increase production intensity to capture economies of scale. This increased production intensity is causing concerns over the environmental, social, and economic sustainability of agro-ecosystems. These concerns are based on observed trends of declining water quality, wildlife habitat loss, reduced biodiversity and the influence of greenhouse gas emissions on climate change. For example, it has been estimated that in North America only 1% of the original tall grass prairie, 19% of the original mixed grass prairie, 16% of the original aspen parkland, and 30% of the original prairie wetlands remain within the prairie landscape (Neave et al., 2000). While these environmental impacts become more apparent, the prairie landscape is becoming increasingly valued by society for the ecological goods and services it provides, including the range of environmental services such as nutrient cycling and the storage and filtering of water, wildlife habitat, and the aesthetic beauty of the landscape itself (Agriculture and Agri-Food Canada, 2000). As a result, there is increasing interest in evaluating and developing policies and programs that provide economic incentives for farmers to adopt agricultural management practices that will increase the provision of ecological goods and services.

The purpose of this article is to examine the role institutions and policy can play in changing the management of prairie landscapes by encouraging or enforcing environmental conservation which will in turn increase ecological goods and services provided by agricultural systems. The article begins with a brief background to the problem and presents a simple economic model to illustrate the forces shaping land management decisions. An overview of institutional arrangements and policy approaches that have been designed to improve ecosystem health and increase economic sustainability is then presented, highlighting the inherent strengths and weaknesses of these approaches. The article then focuses on a particular institutional approach that captures consumers' willingness to pay for ecological goods and services. Specifically, the structure and the advantages and disadvantages of a strategic consumer-producer cooperative are discussed. The concept of consumer-producer cooperatives is presented using an example of the conservation of threatened riparian zones within the 22,000 km^2 internally drained

Wood River watershed, which lies in the mixed-grassland ecoregion in southwestern Saskatchewan. The proposed Wood River cooperative will focus on food production that is certified to be consistent with specific soil, water, and biodiversity conservation objectives.

Landscape Management and Economic Signals

Farmers make investment and land allocation decisions based on a range of internal and external characteristics, including personal ethical and stewardship values and community expectations (McConnell, 1989). However, as most farms are economic enterprises, one of the most important forces driving managers' decisions are economic signals such as input and production costs and output prices. The increasing intensity of agricultural production and declining environmental health of the prairie landscape indicate that allocating agricultural resources primarily to the production of marketable commodities has resulted in altered ecosystem function and a reduction of ecosystem services (Agriculture and Agri-Food Canada, 2000). Agricultural commodities are market goods that can be sold for prices that are attractive to consumers in a cost-competitive retail sector. Environmental goods and services, however, are public goods that are characterized as having no market prices or have market prices that do not reflect their value to society in a meaningful way. The absence of economic signals that indicate the value of public goods, a mechanism to compensate farmers for providing public goods, and/or an effective institutional mechanism to regulate the use of public goods, has resulted in farm management decisions that do not provide a sufficient quantity or quality of these public goods and services from society's perspective.

The economic forces driving land-use decisions can be demonstrated using an economic model (Figure 1). For the sake of illustration, we will consider a simplified case where, at a given time, the land can provide either agricultural products or an environmental or public good (referred to as environmental land in the following discussion) such as water quality. In this case the farmer receives a private marginal benefit (MBp) (reflecting the benefit associated with an incremental or marginal unit of the good or service) from the provision of environmental land. For example, the farmer may experience some satisfaction from providing a level of riparian buffer zone including wildlife-watching benefits and aesthetic benefits. This MBp function is downward sloping, thereby reflecting the fact that the marginal benefit from each additional unit of environmental land decreases as the area or quantity of this land increases.

The private costs of supplying environmental land are described by the marginal costs curve (MCp). This cost largely reflects the opportunity cost of the land to the farmer: this cost represents the benefits (e.g. economic revenue) that could be gained if the marginal unit of land was used for some other activity. For example, within the prairie landscape, the opportunity cost of environmental land would be the revenue that could have been gained from using the land to produce livestock or annual crops. In some landscapes there will be significant quantities of environmental land provided at zero cost (Q_0) because at that time the land has no alternate economic use, or the cost of converting that land to an alternate use exceeds the value. It should be noted that Q_0 is determined by the prevailing economic conditions and the landscape characteristics. For example, Q_0 may decrease

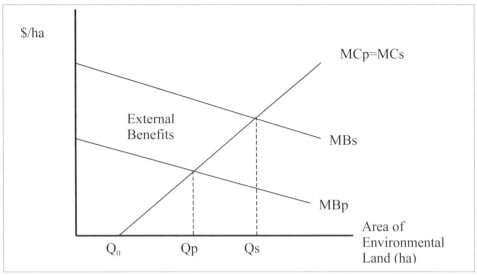

Figure 1. A conceptual economic model for the allocation of land for the provision of environmental goods and services.

with an increase in wheat prices due to an increase in the opportunity cost of environmental land. The private provision of environmental land beyond this point imposes a marginal cost on the farmer, with the marginal cost increasing as the area of environmental land increases and the marginal unit of land that is converted having a higher opportunity cost for agricultural use. Environmental land to the far right of Figure 1 imposes a high MCp since these are relatively productive agricultural lands. All land beyond Q_0 will be allocated to the provision of environmental goods and services only if there is a positive, private net benefit from doing so.

The private market will provide environmental land up to the point where the private marginal benefits (MBp) are just equal to the private marginal cost (MCp), or point Qp in Figure 1. At this point, the private net benefits from land allocation are maximised since the provision of environmental land beyond Qp will decrease private net benefits by the amount that MCp exceeds MBp.

The economic model in Figure 1 indicates that there are external benefits associated with the provision of environmental land that causes the social marginal benefits (MBs) to be greater than the private marginal benefit function (MBp). External benefits are those benefits provided by environmental land, for example, that are captured by society in general but are not captured by the farmer. We assume that there are no external costs associated with the provision of environmental goods and services, such that social marginal costs (MCs) equal MCp, since the extra complication imposed by separate MCs and MCp functions is unnecessary for the discussion. External benefits may include the benefits associated with higher water quality, wildlife habitat and biodiversity, aesthetic benefits, and so on that are enjoyed by society above and beyond the private benefits gained by the farmer. Therefore, Figure 1 reveals that society as a whole attributes a greater value to a unit of environmental land than the private landowner. However, there are no economic signals (e.g., prices) within the market for the farmer to recognize the value of these external benefits. As a result, there is a market failure such that the quantity of environmental land provided by the market (Qp) is less than the quantity of environmental land desired by society (Qs). Societal net welfare would increase (by the distance between MBs and MCs) with every additional unit of environmental land up to Qs. However, for each unit of environmental land provided beyond Qp, there is a net loss of welfare to farmers, since the private costs of providing this land are greater than the private benefits gained (MCp>MBp). Our model suggests that there are land management policies that lead to increases in the provision of environmental land (from Qp to Qs). If these policies were implemented they could increase the quantity and quality of ecological goods and services, and thereby increase net social welfare. The following sections will highlight some institutional and policy alternatives that have been used to accomplish this change in land allocation.

Institutional and Policy Approaches

There has been an extensive body of research on the development of policies and programs to address agro-ecosystem sustainability and the provision of public goods from agricultural landscapes. These policies can increase public good provision through direct command and control regulation, the development of government-administered economic incentives (including taxation or subsidies), and economic incentives provided through institutional change, such as the development of markets. This section will briefly describe the more common policies and institution-based mechanisms developed to increase the provision of public goods within agro-ecosystems.

State-based Mechanisms

It is frequently argued that public goods must be provided publicly rather than privately since the government has the authority to finance these goods using taxes, or to impose fees and to regulate their use (Ostrom, 1998). State-based rights and authority over public goods have been thought necessary to enable the flow of benefits that are not provided through an economic market (e.g., biodiversity, hydrological services and scenic amenities); the equal distribution of benefits from the public goods amongst stakeholders, including farmers, rural and urban residents; and the provision of public good benefits amongst current and future generations (Grafton, 2000). In addition, economies of size may exist for information processing, monitoring, enforcement, and other state management costs. In fact, where markets for privately provided public goods have been established, the state will often retain certain rights over the resource in order to retain control over management. For example, state-based institutional approaches to wildlife issues have been developed at the international level (e.g., The Convention on Biological Diversity, Migratory Bird Convention Act) the national level (e.g., US Endangered Species Act, Canadian Species at Risk Act) and at the regional level (e.g., provincial and state wildlife acts).

While state-based rights are often the default institutional approach to providing public goods, this approach also has problems in democracies. Government management of public goods has, at times, led to resource degradation due to poor information, corruption of enforcement bodies, political objectives that are in conflict with conservation, and a tendency to be influenced by lobbying activities from special interest groups (Grafton, 2000). In addition, government management may have a tendency to be inflexible and can be unresponsive to environmental changes and changes in the resource stock, occasionally resulting in incentives that encourage environmental degradation rather than conservation of public goods.[1]

A common alternative to government regulation and/or direct provision of public goods on agricultural land is to provide economic incentives for provision in the form of subsidies or disincentives to degradation through taxes. In the context of privately owned farm land, the subsidy approach, or "Provider Gets Principle" (Hanley et al., 1998), involves the identification of a level of public good provision as a policy objective (e.g. a target of X kilometres of riparian area being vegetated). The government then directs public funds to the providers according to their marginal opportunity cost of supply. This compensation for opportunity costs is necessary because landowners cannot be forced to supply these public goods on their private land. In general, standard payment rates are offered to farmers who will voluntarily agree to a set of management alternatives that will directly or indirectly supply the targeted public goods or services. Examples of this mechanism include the North American Waterfowl Management Plan that was funded by the Canadian and American governments to increase waterfowl habitat (Riemer, this issue) and the Conservation Reserve Program funded by the American government to provide a range of public goods and services.

While the state-funded provision of public goods is common, an efficient or optimal level of these goods and services may not be provided because of difficulties associated with estimating the marginal benefits of public goods. The heterogeneity of farms across a region results in very different opportunity costs of public-good provision, making it difficult to design programs that are equitable for a majority of the farms (Hanley et al., 1998). Furthermore, lobbying by political, agricultural, consumer or environmental groups may influence state programs in ways that do not encourage the provision of public goods.[2] In addition, government incentive programs often require significant amounts of funding to be made available over relatively long time frames in order to secure the long-term provision of public goods. Ongoing changes in governments and political objectives hamper the ability of these programs to provide public goods in a sustainable manner.

Market-based Mechanisms

An alternative to government provision of public goods is the establishment of institutions that facilitate the development of economic incentives for public-good provision through a market. A market approach to public-good provision has been identified by Hanley et al. (1998) as the "Beneficiaries Pay Principle," whereby individuals who benefit from the good pay a marginal value-based fee to the provider of the good. The fee payment compensates the provider of the public good for their opportunity cost of provision. The interaction of the marginal willingness to pay (demand) of the consumer and the cost of supply (opportunity cost) of the provider can result in an efficient level of public-good provision.

One example of a market solution to the provision of public goods from agricultural land is privately provided hunting, or fee hunting, where landowners are allowed to charge a fee to hunters who hunt on their land. While the state retains ownership of the game, the landowner manages the habitat and the hunting. The fee that hunters and fishers pay to the landowner serves as an economic incentive for the provision of habitat for the game species. Hunters and fishers may be willing to pay higher fees to access areas that provide greater probability of success, which provides an incentive for the landowner to invest in the provision of a stock of premium wildlife habitat on their land. Further, the increased area and improved quality of habitat will potentially provide a number of other benefits, including increased numbers of non-game species (increased biodiversity) and other ecological goods and services.

While the fee-hunting type approach may improve the provision of certain goods and services, it is most appropriate for those goods and services that have use values, and in particular, consumptive use values such as wildlife game species. In contrast, goods and services that have value to society based primarily on non-consumptive use or non-use values, such as hydrological services and many types of non-game wildlife, may not be well served by this approach. There will be no incentive to make investments in the provision of these non-market goods and services, and in certain cases, this market approach may introduce incentives to eliminate species that compete directly or indirectly with the game species. An exception to this occurs when non-consumptive use values can be captured through eco-tourism markets. For example, within Saskatchewan there is a network of "Holiday Farms" in which consumers pay the farmer a fee to experience the range of agricultural and natural benefits associated with the rural setting.

Another example of a market-based approach is based on the identification of specific environmental goods and services as attributes of market goods. Growing evidence shows that consumers care about the environment and are willing to pay a higher price for a product or process that generates less environmental harm (e.g., Shrum et al., 1995). For example, Matoo and Singh (1994) report that up to 80% of Canadian consumers expressed a willingness to pay extra for environmentally friendly commodities. Environmental labels, or eco-labels, have been the most common way to ascribe environmental goods and services to a particular market good. The eco-label provides a financial incentive for the producer to adopt production strategies that are less environmentally damaging or more sustainable. Well known labelling programs include the "dolphin-friendly tuna" label, which identifies tuna that has been harvested using nets that minimize the number of incidentally killed dolphins; and labelling programs for sustainable forestry practices (like the Forest Stewardship Council, Sustainable Forestry Initiative, Eba'a et al., 2002). Other environmental labels have been developed to identify products that are linked to a range of environmental goods and services (Table 1: Eco-Labels, 2002).

Environmental labels enable consumers to express their preferences concerning public goods through the market. The advantage of this approach is that it provides a mechanism to economically quantify the value of the public goods to society. In the case of food, the public-good attributes can include the level of environmental goods and services conserved, enhanced, or created during the production process. A challenge in any environmental labelling scheme is the establishment of a credible verification process, which can add a significant cost. Yet, for consumers to be willing to pay a premium for an environmentally labelled product they need to be assured that the product is providing those attributes promised on the label. The reputation of the producer may be sufficient to engender trust in specific cases of such public-good provision. However, where a premium is available for products

Table 1. Examples of Eco-labels, Basic Structure and Identified Goals.

Name	Structure	Goals
Bird-friendly label[a]	Certified by arms-length body Smithsonian Migratory Bird Centre	Shade grown coffee Bird habitat on plantations and riparian protection.
Community-supported agriculture[d]	Consumer-producer food contracts	Known food growing conditions Advance orders help meet costs Trading food for labour
Food Alliance[c]	Eligibility guidelines Production and land conservation plans available	Improve food quality Protect environment Provide "organic" food Conserve soil and water Ensure safe and fair working conditions Family and locally raised land conservation plan prepared
Foodtrust of Prince Edward Island Ltd.[b]	Branding food products with audit for marketing Voluntary producer-retailer partnerships	Economic, environmental, and social sustainability Develop markets
Salmon Safe[a]	Certified by arms-length body Pacific Rivers Council	Protection for salmon spawning sites

Sources: [a] Eco-Lablels, 2002; [b] Davey, 2001; [c] Hamilton, 2001; [d] Craig, 1994.

with the environmental label, there is an economic incentive for the producer to label their products without actually providing the claimed environmental attributes. For example, Iyer and Banerjee (1993) report that nearly one-third of all green advertising claims were nothing more than vague claims or attempts to align products with the growing green movement.

Consumer-Producer Cooperatives

An alternative mechanism to ascribe a price premium to market goods, such as food, that provide specific environmental or public-good attributes is through a cooperative relationship between producers and consumers. Environmental labels and consumer-producer cooperatives stimulate change in the management of the prairie landscape through similar mechanisms. Both approaches enable consumers to express their preferences for public good attributes of agricultural commodities, thereby providing an incentive for farmers to increase their investment in the public-good provision. The cooperative could involve a specific group of producers who provide food products that have some environmental attribute and a group of consumers who are willing to pay a premium for the knowledge that the environmental attribute is provided. This type of cooperative has characteristics of a new generation cooperative including: being more specialized and interlinked up and down the value chain; having the capacity to address specific consumer markets; stressing quality, niches and special characteristics of crops and animals; fostering tight, mutually beneficial contractual relationships with members; and providing investment into human, social and physical capital (Fairbairn, 2003). A number of different forms of food cooperatives have developed that include in their objectives such issues as environmental sustainability or the support of ecologically responsible producers.[3] However, many of these are primarily consumer cooperatives, whereas the proposed Wood River Cooperative will have a stronger direct linkage between producers and specific consumers.

The cooperative proposed here also has strong similarities with community-supported agricultural groups (CSA). CSA has been described as a partnership of mutual commitment between a farm and a community of supporters that provides a direct link between the production and consumption of food (O'Hara and Stagl, 2002). In most cases the aim of these groups is to provide consumers with healthy, locally grown food, and at the same time to revitalize local food economies, enhance local food security, protect the environment, and preserve small-scale, family-farm food production. In evaluating two CSA groups in the United States, O'Hara and Stagl (2002) believe that environmental concerns and interest in the local economy are strong motivational factors for joining and remaining a member of a CSA.

There are a number of specific advantages to consumer-producer cooperatives in the context of providing ecological goods and services. It should be highlighted that this type of cooperative would typically involve a relatively small number of producers and consumers. To meet these needs, we are using a case study of the proposed Wood River Cooperative in Saskatchewan. The Wood River Cooperative is expected to initially involve 30 to 50 families as consumers and to operate on portions of land bordering the Wood River belonging to 15 to 20 farmers. As a result, the focus can be on fairly specific local issues such as the sustainable management of a threatened ecosystem or the conservation of habitat of a specific local wildlife species. The cooperative could also be broadened to include a focus on the particular economic and/or social sustainability issues of a local area or a larger region. The social capital, including consumer trust that may build between producers and consumers within the cooperative, would significantly decrease the need for expensive third-party verification as required in labelling initiatives. In addition, the cooperative could be quite flexible, enabling relatively rapid changes in the bundle of food products and environmental attributes provided, based on changes in consumer preferences, producer ability, food supply-chain opportunities, and environmental dynamics. Finally, consumer-producer cooperatives make explicit use of human capital, including site-specific knowledge about public goods held by individual producers, such as locations of wildlife populations and habitat use, local hydrological characteristics, and the appropriateness and effectiveness of management alternatives.

While there are some significant advantages to consumer-producer cooperatives in providing public goods within the prairie landscape, small market size may, to some extent, limit the regional benefits of this institution. For example, very small consumer-producer cooperatives will be able to effectively address only local environmental issues, and administration and transportation costs may hamper development. In addition, there may be a role for the government in decreasing costs of cooperative creation and operation. As such, consumer-producer cooperatives could be an important component of a landscape-scale strategy aimed at increasing ecosystem health and providing a range of ecological goods and services to society.

The Wood River Consumer-Producer Cooperative

The Wood River Ecosystem

The 22,000 km² Wood River watershed is a self-contained basin in Saskatchewan and is located immediately north of the Missouri-Mississippi (Gulf of Mexico)

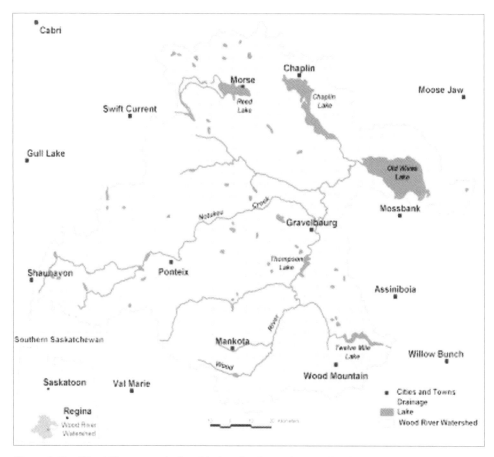

Figure 2. The Wood River watershed and its location in southern Saskatchewan.

drainage basin as representedin Figure 2, but within the Saskatchewan-Nelson (Hudson Bay) drainage basin (Fung et al., 1999). Since the establishment of an agricultural economy, farming has undergone profound changes in the region due to climatic conditions, and socio-economic and technological changes. Agricultural production in the Wood River watershed has become highly specialized in the production of a few commodities. Land-use statistics for crop district 3AN, which contains the Wood River watershed, indicate that in 1999, approximately 53% (218,000 ha) of the land in annual crop was dedicated to spring wheat with another 28% (114,000 ha) in durum wheat. Other annual crops include barley (18,442 ha), oats (8621 ha), canola (32,849 ha), and flax (10,939 ha) and some specialty crops (e.g., lentils) (Saskatchewan Agriculture and Food, 2001). In addition, there were 36,800 beef cows (excluding bulls, steers, and heifers), 200 milk cows, 4,900 pigs, and 800 sheep. According to land cover interpretation from 1986 satellite imagery, only about 10% of watershed is native prairie, and this land is located primarily where glacial moraine and the slopes along the Wood Mountain upland made cultivation difficult.

Land cover changes and agricultural development have altered the biodiversity of the watershed. While some species have increased since settlement, such as several species of geese that stop in the area during migration, many other species have declined. For example, the plains grizzly (*Ursus arctos horribilis*), prairie wolf (*Canis latrans*), and sage grouse (*Centrocercus urophasianus*) have been extirpated, the burrowing owl (*Athene cunicularia*) continues to decline and populations of many other animals and plants within the watershed have diminished.

The removal of prairie sod, an effective water-holding sponge (De Boer, 1994) and widespread draining of permanent and semi-permanent wetlands (Ignatiuk and Duncan, 1995), has resulted in significant changes to the hydrology of the region. In addition, the removal of vegetative cover from the river's slopes has led to wind and water erosion resulting in the transport of soil and agro-chemicals into the river (Holm, 2004). On each of the 98 farms adjacent to the river in 2001 (Bradshaw and McIver, 2001), a small proportion of the land directly impacts the river since a very small area lies within the riparian zone, yet along the river's length, erosion is common. While seepage springs were once commonplace in the region (Joan Richmond, pers. comm.), water quality and quantity is now a major concern and has been identified by the local community as a contributor to rural depopulation (V. Thibault, pers. comm.). Preliminary data from a recent study indicate that phosphorous in the Wood River exists at many times the levels recommended for consumption by humans and livestock, or for recreation (J. Holm, 2004). This evidence suggests that within this agricultural landscape, a market failure has resulted in an under-provision of such public goods as wildlife habitat/biodiversity, hydrological services, and water quality.

The Policy Context

In response to these environmental and economic sustainability issues, a number of state-based strategies have been implemented in the Wood River watershed, including government legislation and government-funded subsidy and taxation programs. The legislative approach is well represented in Saskatchewan with no less than seven Acts designed to protect the quality of surface water. These include the Conservation and Development, Groundwater Conservation, Irrigation, Water Corporation, Watershed Association Environmental Management and Protection acts, and most recently the Saskatchewan Watershed Authority Act. As discussed earlier, these state-based legislative approaches may not be sufficiently sensitive to specific local demands. While legislation has the potential of affecting all of a landscape, its ability to protect the selective use of permanent vegetative cover on slopes bordering the Wood River is weak. For instance, a buffer strip around water bodies is widely recommended and in some cases even required by law (Davey, 2001). A 10–30 metre buffer guideline is recommended in Saskatchewan (Huel et al., 2000) but not commonly implemented. Regardless of the legislation or policy, however, enforcement is difficult along the Wood River. Furthermore, legislation and guidelines appear to be preventing local landowners from finding and developing viable alternative management approaches.

Government-funded incentive programs, consistent with producer-gets-principle initiatives discussed earlier, have also been implemented in Saskatchewan and within the Wood River watershed. For example, the Saskatchewan Government's permanent cover program was designed to address soil erosion, but it has been costly to implement and as a result has had only a minor impact at the landscape level. Further, programs of this type have historically been ephemeral due to changing government administrations and priorities. Most importantly, specific conservation needs for maintaining even basic ecosystem functions such as vegetation buffers for watershed protection far outstrips funds available for incentives in Saskatchewan and likely elsewhere. Balmford et al. (2003) find that the cost of protected areas in reserves and agri-environment programs in the UK ranged from US $15,000 to $50,000 per km^2 per year.

An alternative program that falls within the producer-gets-principle category of policies is funded stewardship. Funded stewardship, such as is practiced in Nature Saskatchewan's Operation Burrowing Owl program or by the Saskatchewan Wildlife Federation, is initiated by special interest groups providing financial compensation to specific farmers who adopt management practices that ensure the provision of some environmental amenity (e.g., burrowing owl habitat). Another example of privately funded stewardship are 10-year hay establishment agreements with landowners through the combined efforts of the Saskatchewan Wetland Conservation Corporation (now the Saskatchewan Watershed Authority) and the Notukeu Wildlife Federation. This type of initiative has been found to be locally beneficial for increasing environmental capital. However, with the vast majority of the agricultural landscape being privately owned and allocated to the production of marketable commodities, a systems perspective would suggest that environmental protection should be placed squarely within a production system, not outside of it. Recently, as part of the development of the Wood River cooperative, "seed-funding" was provided to five landowners to plant a similar permanent buffer strip on their own and neighbours' lands along the Wood River for hay production for a five-year term (Yeager and Annand, 2002).

The Consumer-Producer Cooperative

The proposed consumer-producer cooperative is a market- and system-based approach to addressing the market failure that is compromising environmental and economic sustainability of the Wood River agro-ecosystem. The cooperative is currently in the development stages and is being initiated as a collaboration between a sustainability-oriented citizenry in and outside of the watershed, particularly between researchers at the Centre for Studies in Agriculture, Law, and the Environment (CSALE) at the University of Saskatchewan and farmers in the Wood River area. As Schmutz (2003) notes, the initiative received its initial impetus from Saskatchewan's Important Bird Areas (IBA) Program.[4] During the community conservation planning phase of the IBA program, it became evident that people's health and sustainability in the region were as threatened by poor water quality as were the birds. Conceptually, the initiative combined diverse ideas and new approaches in the areas of environment, sustainability, and community-based action, notably those of community-supported agriculture (discussed earlier) and locally, informally organized sustainable agriculture initiatives, such as those Jackson and Jackson (2002) propose. Currently, research is underway to ascertain where action is most needed and which measures might best serve the holistic goals. Also, consumers in the city of Saskatoon are being canvassed for their participation in the cooperative. Saskatoon is a city of approximately 200,000 located 200 km north of the watershed (Figure 2).

The principle goal in the development of this cooperative is to construct reciprocal linkages between farmers and urban consumers to achieve a quality of life for the wider local community and to maintain ecosystem function. It is proposed that the cooperative will facilitate an expansion of stewardship with human and manufactured capital that arises from within the food system. The cooperative should have appropriate checks and balances to protect the Wood River's riparian ecosystem, provide a personally and economically rewarding livelihood for farmers and employees of the cooperative, provide fresh and high quality food for consumers, and engender a feeling of community and satisfaction through environmental protection. For logistical reasons, farmers, ranchers, and food processors should be in close proximity, and the consumers should be close together to ensure an efficient delivery system, but the two groups themselves can be connected through long-distance transport.

Environmental Goals

The principal environmental management objective of the food cooperative will be riparian zone conservation to retard or prevent wind or water from transporting silt and agro-chemicals to the adjacent Wood River or its tributaries. Permanent cover in the form of alfalfa or non-invasive grasses will be planted in the sloping riparian zone for hay production or grazing. Spin-off benefits can be expected at various levels.

Landscape Level:

• Enhanced snow capture in winter and reduced evaporation in summer allow land bordering the river to act as a sponge that retains water for slow release.

• Corridors for movement of animals that are particularly vulnerable in open fields will improve (e.g., tiger salamander (*Ambystoma tigrinum*) and garter snake (*Thamnophis* sp). This will be particularly important in the extensively cultivated northern portion of the Wood River Plain.

• Improved escape cover will allow species to use fields (e.g. at night) but be able to seek refuge periodically (by day).

Aquatic Communities:

• Reduced soil erosion into the river will allow the river to retain a mix of mud and rocky substrate for fish habitat.

• Reduced nutrient loads being transported directly into the river decrease algal blooms which deplete oxygen when algae die.

• Pesticides loads being transported directly into the river will be reduced.

In addition, the Wood River and its tributaries are the source of water for two Important Bird Areas that have earned several international designations: Old Wives and Chaplin lakes. The watershed supports species-at-risk, e.g. burrowing owls (*Athene cunicularia*) and piping plover (*Charadrius melodus*). Therefore, the objectives of the cooperative will help to conserve critical biodiversity, an important component of the ecological goods and services provided by the area. Nature viewing opportunities will be organized for members of the cooperative and this is expected to maintain interest and participation.

Cooperative Structure

The development of the cooperative requires the participation of both landowners in the Wood River watershed and consumers. Participation and success will be enhanced by involving women in all aspects of cooperative development (W. Siemens, pers. comm.; Krug, 1999). Locally, we are aware of considerable

disenchantment with existing corporate structures in the food system, and this disenchantment has lead to a proliferation of direct marketing and other local food
alternatives. We believe that a cooperative structure can provide a land-to-
consumer link, facilitate trust in production methods, provide opportunities for
inspecting changes on site, and thus allow for a renewal in the context of economic, social, and ecological realities. A cooperative structure also provides flexi-
bility in food inspection requirements. Gertler (2003) states "Cooperatives offer
advantages as rooted, socially embedded, patient capital and as organizations that
promote partnerships, coordinated action, and capacity building."

The primary land-management focus of the cooperative is conservation of the
riparian zone of the Wood River through the establishment of a vegetated buffer
of appropriate width for land slope, erosion potential and river meanders (e.g.,
Wolf et al., 2004). Livestock use of the riparian zone will incorporate managed
grazing using electric fencing and remote watering systems to avoid excessive tram-
pling. Land for buffer strips will be rented from participating landowners. An
employee of the cooperative will execute day-to-day management including over-
seeing contractual agreements with local people for specific services where appropriate (e.g., slaughter, processing, and food deliveries). It is necessary for the cooperative to assume some level of management responsibility, as farming in the
region has become specialized toward few commodities. Food production strategies in the riparian zone will be moderately diverse and designed based on opportunities provided by the land, consumer desire, food production, processing, delivery, and storage. A management team will be struck that includes consumer and
landowner cooperative members and "armslength" representatives from environmental and sustainable agriculture groups. This team will need to ensure that the
founding principles of the cooperative are maintained, that monitoring is appropriate, and that adaptations, where needed, are consistent with the principles. The
initial costs of seeding the selected riparian land base will be at least partially offset
with financial support from assistance programs such as the Government of
Canada Green Cover Program.[5] The group of urban consumers will be enlisted by
focusing initially on members of environment and nature organizations. Since the
organizers of the Wood River cooperative are based in Saskatoon, for logistical reasons, the initial group of consumers will be enlisted from Saskatoon, with potential
to establish linkages with consumers in other urban areas in the future.

The initial product goal of the cooperative is to provide grass-fed beef and
chicken produced without growth hormones and using a minimum of other phar-
maceuticals (e.g., vaccination). Other attributes of the food product will include
on-the-hoof transport of cattle where possible, local slaughter, and meat aged to
achieve maximum quality. High-quality food will be an important feature making
the cooperative attractive to the consumer and to maintain participation. Other
food products may be added and the complexity of products will increase as the
cooperative matures. Products likely to be added include a mix of grains and fruit
sold as granola, root crops, and vegetables in season. These subsequent products
will be produced using management that adheres to the same conservation stan-
dards as the initial cooperative products.

The cooperative is expected to build a level of consumer trust in the food products. This trust will be fostered through transparency and information sharing
(e.g., web site, newsletter, field trips, product labeling, and farm management

plans). A number of bed-and-breakfasts operate in the region and tours can high-light farm visits, bird watching, local parks and recreation areas, and local heritage festivals.

There are significant advantages offered by the model consumer-producer coop-erative discussed here. It is important to note, however, that it is quite uncertain whether this type of consumer-producer cooperative will ever constitute a signifi-cant part of future food markets. Thus, members of the Wood River Cooperative need to be satisfied that their impact will be significant on a local scale, but possi-bly less important at a broader regional, provincial, or national scale. Prairie ecosys-tems may be sufficiently resilient to cope with the stresses imposed by convention-al production in many regions, while some specific areas may be extremely vulner-able to these production practices. Initiatives such as the Wood River Cooperative may be best viewed as part of a food production mosaic, including conventional and niche-group inspired production. The proposed system's greatest strength may lie in its ability engage consumers seeking not only quality but meaning in food; it presents a chance to opt out of corporate culture and align one's purchas-ing power with ones social, environmental, and ethical values.

Conclusion

Concern over the environmental and economic sustainability of prairie agri-cultural systems has influenced the development of a number of policy and insti-tutional approaches aimed at modifying the management of the prairie landscape. In general, these approaches alter the economic signals received by farmers in order to correct the market failure that results in low investment in public-good provision by farmers. The development of strategic consumer-producer coopera-tives that have specific environmental objectives can play a part in increasing the sustainability of agricultural systems. These cooperatives enable consumers to express their preferences for a range of public goods, while providing a premium to farmers that can serve as an economic incentive to increase investment in public-good provision. An important advantage of this approach is that it makes explicit use of extant human capital. In this context, human capital refers to the local knowledge about local public goods held by individuals. Human capital may include knowledge, such as locations of wildlife populations, local hydrological characteristics, appropriateness of different management practices, and a range of other system knowledge. This site-specific knowledge, often developed by trial-and-error or learning-by-doing methods, and contained in local communities, is very useful in the management of the local resource base and the efficient provision of public goods. It has been shown that institutional approaches that ignore or degrade this site-specific knowledge can encourage neglect and increase exploita-tion of the natural environment (Agee and Crocker, 1998). The institutional struc-ture of the Wood Mountain Cooperative enables both consumers and producers to proactively respond to their concerns about the sustainability of agricultural ecosys-tems and, as such, have the potential to contribute to the sustainable management of the prairie landscape.

Notes

1. As correctly noted by one reviewer, these concerns may, at times, be more of an issue of scale than an issue of government involvement. Depending on the issue of concern, a local government may be very flexible and able to respond quickly to changes in the resource and the need for a modified program.

2. For a more complete discussion of lobbying activity and "rent-seeking behaviour" in agricultural systems see Schmitz et al. (2002).

3. See the University of Victoria BC Institute for Co-operative Studies (http://web.uvic.ca/bcics) and the University of Wisconsin Center for Co-operatives websites (http://www.wisc.edu/uwcc) for extensive lists of links to existing food cooperatives.

4. See http://www.ibacanada.net for a complete description of the Important Bird Areas Program.

5. See http://www.agr.gc.ca/env/greencover for a complete description of the Canada Green Cover Program.

References

Agee, M.D. and T.D. Crocker. 1998. "Environmental Change, Institutions, and Human Capital." In Loehman, E.T. and D. M Kilgour (eds.), *Designing Institutions for Environmental and Resource Management.* Cheltenham, UK: Edward Elgar Press.

Agriculture and Agri-Food Canada. 2000. *Prairie Agricultural Landscapes: A Land Resource Review.* Prairie Farm Rehabilitation Administration, Minister of Public Works and Government Services, www.agr.ca/pfra/pub/pallande.pdf

Balmford, A.K.J., G.S. Blyth, A. James and V. Kapos. 2003. "Global Variation in Terrestrial Conservation Costs, Conservation Benefits, and Unmet Conservation Needs." *Proceedings of the National Academy of Sciences* 100: 1046–1050.

Bradshaw, A.L. and S.E. McIver. 2001. *Wood River Riparian Project: A Final Report.* Saskatoon and Gravelbourg, SK: Centre for Studies in Agriculture, Law and Environment, University of Saskatchewan, Saskatoon, and PFRA.

Craig, J.G. 1994. "Japanese Consumer Co-ops Create Caring Community, Sound Environment," *Alternatives* 20, no. 2: 11.

Davey, S. 2001. "Foodtrust—Soil Red, Ocean Blue." *The Atlantic Co-operator* (November).

De Boer, D.H. 1994. Lake *Sediments as Indicators of Recent Erosional Events in an Agricultural Basin on the Canadian Prairies: Variability in Stream Erosion and Transport.* Proceedings of a Canberra Symposium, IAHS Publ. No. 224.

Eba'a Atyi, R. Simula and M. Simula. 2002. "Does Green Timber Make a Difference: A Global Status Report on Forest Certification Schemes," *Conservation in Practice* 3, no. 4: 26–27.

Eco-Labels. 2002. website. http://www.ecolabels.org/home.cfm

Fairbairn, B. 2003. *The Role of Farmers in the Future Economy.* Saskatoon: Centre for the Study of Co-operatives, University of Saskatchewan.

Fung, K., B. Barry, and M. Wilson (eds.). 1999. *Atlas of Saskatchewan.* Saskatoon: University of Saskatchewan.

Gertler, M.E. 2003. "Synergy and Strategic Advantage," *Journal of Cooperatives.*

Grafton, Q. 2000. "Governance of the Commons: A Role for the State?" *Land Economics* 76, no. 4: 504–517.

Hamilton, N.D. 2001. "Putting a Face on Our Food: How State and Local Food Policies can Promote the New Agriculture." Pages K-4-1–K-4-35, presented at the American Agricultural Law Association, at Colorado Springs, CO.

Hanley, N.H. Kirkpatrick, I. Simpson and D. Oglethorpe. 1998. "Principles for the Provision of Public Goods from Agriculture: Modeling Moorland Conservation in Scotland," *Land Economics* 74, no. 1: 102–113.

Holm, J. 2004. "The Impact of Land Use on Water Quality in the Wood River, Saskatchewan." Masters thesis, Department of Environmental Engineering, University of Saskatchewan.

Huel, D., T. Harrison, A. Foster and N. Parwana. 2000. *Managing Saskatchewan Wetlands: A Landowner's Guide.* Regina: Saskatchewan Wetland Conservation Corporation.

Ignatiuk, J. and D.C. Duncan. 1995. "Wetland Loss in Aspen Parkland of Saskatchewan," *Blue Jay* 53: 129–133.

Iyer, E. and B. Banerjee. 1993. "Anatomy of Green Advertising." Pp. 292–298 in L. McAlister and M. Rothschild (eds.), *Advances in Consumer Research.* Provo, UT: Association for Consumer Research.

Jackson, D.L. and L.L. Jackson (eds.). 2002. *The Farm as Natural Habitat: Reconnecting Food Systems with Ecosystems.* Washington, DC: Island Press.

Krug, K.L., 1999. "Canadian Rural Women Reconstructing Agriculture." Pp. 167–173 in M. Koc, R. MacRae, L.J.A. Mougeot and J. Welsh (eds.), *For Hunger-proof Cities: Sustainable Urban Food Systems.* Ottawa: International Development Research Centre.

McConnell, K.E., 1989. "The Optimal Quantity of Land in Agriculture," *Northeast Journal of Agricultural and Resource Economics* (October): 63–72.

Matoo, A. and H.V. Singh. 1994. "Eco-labelling Policy Considerations," *Kyklos* 47: 53–65.

Neave, P.E., T. Weins and T. Riche. 2000. "Availability of Wildlife Habitat on Farmland." Pp. 145–158 in. T. McRaie, C.A.S. Smith and L.J. Gregorich (eds.), *Environmental Sustainability of Canadian Agriculture: A Report of the Agri-Environmental Indicator Project*. Ottawa: Agriculture and Agri-Food Canada.

O'Hara, S.U. and S. Stagl. 2002. "Endogenous Preferences and Sustainable Development," *Journal of Socio-Economics* 31: 511–527.

Ostrom, E. 1998. "The Institutional Analysis and Development Approach." In Loehman, E.T and D.M. Kilgour (eds.), *Designing Institutions for Environmental and Resource Management*. Cheltenham, UK:Edward Elgar Press.

Saskatchewan Agriculture and Food. 2001. *Agricultural Statistics 1999*. Regina: Statistics Branch, Saskatchewan Agriculture and Food.

Schmitz, A., H. Furtan and K. Baylis. 2002. *Agricultural Policy, Agribusiness, and Rent Seeking Behaviour*. Toronto: University of Toronto Press.

Schmutz, J.K. 2003. "Important Bird Areas of Saskatchewan," *Blue Jay* 61: 193–198.

Shrum, L., J. McCarty and T. Lowrey. 1995. "Buyer Characteristics of the Green Consumer and Their Implications for Advertising Strategy," *Journal of Advertising* 24: 71–82.

Wolf, T. M., A. J. Cessna, B.C. Caldwell and J.L. Pederson. 2004. "Riparian Vegetation Reduces Spray Drift Deposition into Water Bodies." In A. . Thomas (ed.), *Field Boundary Habitats: Implications for Weed, Insect, and Disease Management. Topics in Canadian Weed Science*, Volume 1. Sainte-Anne-de-Bellevue, QC: Canadian Weed Science Society.

Yeager, Penny and Mel Annand. 2002. "Wood River Riparian Protection Project Forage Partnership." Unpublished report, Centre for Studies in Agriculture, Law and Environment, University of Saskatchewan.

CHAPTER 6

Mutual Trust in Community-Based Ecosystem Management: Early Insights from the Frenchman River Biodiversity Project

Glenn C. Sutter, Diane J. F. Martz, Jean Lauriault, Robert A. Sissons and Jana Berman

ABSTRACT. This article examines some of the challenges associated with transforming a "top-down" idea into a "bottom-up" research project, giving special attention to the importance of fostering mutual trust as a form of social capital. The insights we discuss are based on discussions and events that gave rise to the Frenchman River Biodiversity Project (FRBP), an interdisciplinary study of aquatic biodiversity and social sustainability in southwest Saskatchewan. The impetus for the FRBP originated at the national level, yet the governance structure of the project is biased towards local interests. Our aim here is to review how this structure and other aspects of the project developed, as a case study in community-based ecosystem management (CBEM).

SOMMAIRE. Cet article examine quelques-unes des difficultés qui s'opposent à transformer une idée "descendante" en projet de recherche "ascendant," et considère plus particulièrement l'importance d'inspirer une confiance mutuelle comme forme de capital social. Les aperçus que nous discutons sont basés sur des discussions et événements qui ont donné naissance au Projet Biodiversité de Frenchman River (FRBP), une étude interdisciplinaire de la diversité aquatique et de la soutenabilité sociale du sud-ouest de la Saskatchewan. L'impulsion pour ce projet s'est déclenchée au niveau national, mais sa structure d'autorité est orientée vers les intérêts locaux. Notre but est de décrire le développement de cette structure et d'autres aspects du projet, et d'accomplir ainsi une étude de cas sur la gestion communautaire d'un écosystème.

Introduction

Some features of the current Canadian prairie landscape, including railways, highways, large dams, and national parks, are the consequence of "top-down" planning by distant institutions. "Top-down" planning is necessary for sustainable development because some critical resources, such as freshwater, are held in common. But "top-down" strategies usually affect a relatively small area for each dollar invested (Riemer, this volume), and if they spark conflict or result in societal polarization, they may reduce the internal social capital that communities require to respond to changing conditions (Flora, 1995; Diaz and Nelson, this issue). At the other extreme, "bottom-up" processes can affect a large area if many people contribute to them. Based on the positive response of waterfowl populations to recent, economy-driven changes in farming practices (Riemer, this volume), "bottom-up" action by individual producers is the most cost-effective way to alter some portions of the Canadian prairie landscape. Presumably, "bottom-up" processes are also more likely to have a positive impact on the social capital of rural communities.

New approaches to community planning emphasize local people and local institutions as participants in community planning processes. These approaches, including community-based ecosystem management (CBEM) strategies, are founded on the premise that development will be more successful if it "starts from a base of local resources and involves popular participation in the design and implementation of development action" (Ray, 1999: 524). Another assumption is that people will more readily commit to a management plan and that these commitments will be longer lasting, when individuals or small groups develop initiatives on their own and the decisions they make lead to results that are tangible and local (Schindler and Cheek, 1999). In addition, where environmental monitoring is the goal, efforts that are assisted or conducted entirely by community volunteers, which can provide highly valuable quantitative data for scientific and official purposes (Martel et al., 2001; Engel and Voshell, 2002), may be facilitated when a bottom-up approach and CBEM strategies are involved.

The aim of a CBEM project is to conserve biodiversity and to monitor and maintain ecosystem functions in the context of local land-use practices. To be successful, CBEM projects must be accessible, administratively transparent, socially inclusive, adaptive to changing conditions and circumstances, and involve mutual learning (Gray et al., 2001; Moote et al., 2001). In order to make good decisions that consider issues of human and ecological well-being, competence, fairness, strength, efficiency, and opportunities to learn (Dietz, 2003), the agencies involved need to work closely with local communities and individuals, allowing both groups to contribute their knowledge, information, experience, and energy to management processes (Ack et al., 2001). Biodiversity is the focus because the number and condition of species, where they live, and how people relate to them can shed light on the resilience of social and ecological systems. Biodiversity also provides a useful foundation for talking about related concepts, such as stewardship and ecosystem health. Stewardship is often a tangible concept for a lay audience because it assumes that people are an integral part of their ecosystems, whether they are hunters after game or producers in agricultural systems. Similarly, the concept of ecosystem health (Rapport, 2000) places biodiversity in a broader context, highlighting its role in water regulation and other ecosystem services (Constanza et al., 1997).

CBEM projects aimed at whole watersheds are likely to involve numerous stakeholders with competing interests and different world views (Webler and Tuler, 2001; Johnson et al., 2003). A biologist might say that the health of a river is reflected in the plants and animals that live in the water and along the shoreline. Ranchers, farmers, and local communities may be more interested in the capacity of a river to provide reliable supplies of good-quality water for crops, livestock, and human use. Shoreline residents may be more concerned about the clarity of the water, the aesthetics of shoreline vegetation, or the presence of nuisance species. A central challenge in CBEM research is to document, clarify, and compare these points of view and how they affect land management practices and the social, economic, and environmental dimensions of sustainability.

On the Canadian pPairies, where rural people tend to be wary of "outsiders" (Diaz and Nelson, this volume), one of the biggest challenges associated with CBEM projects is the need to foster mutual trust, a critical feature of social capital (Putnam, 2001). For projects to succeed from conservation and management perspectives, mutual trust needs to exist between outside organizations and local communities, within the regional planning team, and between local proponents and their home communities. In this article, we examine the importance of trust in all of these dimensions, using the Frenchman River Biodiversity Project (FRBP) as a case in point. First, we report on the evolution of a project structure that provides for liaison with relevant agencies and gives local residents a measure of control over project activities. We also describe steps that were taken to accommodate requests for land-owner confidentiality in the collection and management of sensitive biological information. Then, in light of concerns voiced by project scientists, we report on feedback received from journal editors in response to an enquiry about the repercussions of a confidentiality agreement on the publication of scientific findings. In the final section, we examine the importance of fostering trust in local participants.

Project Overview

The FRBP is an interdisciplinary and multi-agency research project that began in 2003 and will run until 2006. A joint undertaking of the Royal Saskatchewan Museum and the Canadian Museum of Nature, it stems from a working partnership involving local residents, university researchers, and representatives from provincial and federal government departments. Based on needs and principles that were defined by an interim Steering Committee (Appendix 1), the aim of the project is to assess the health of the Frenchman River and the sustainability of local activities by studying the aquatic biodiversity of the river and social interactions that affect the watershed. The research goals are to compare the water chemistry and aquatic biodiversity associated with a range of land uses and management practices, and to assess local knowledge and perspectives about the health of the river. A central management goal is to equip local residents with a community-based mechanism they can use to keep the river healthy. Other desired outcomes identified by the Steering Committee include tools and processes that local stakeholders can use to monitor the health of their watershed; increased awareness of the value of aquatic biodiversity as a measure of ecosystem health, including information about how various species and their interactions contribute to specific ecosystem functions and services; information about local aquatic biodiversity "hot-spots" and indicator species; insights about where, how, and when various stewardship practices should be applied; and local opportunities for professional development and employment.

The Frenchman River watershed covers over 700,000 ha of southwest Saskatchewan and supports a population of about 2000 people, including the towns of Val Marie and Eastend (Plate 1). Ranching is the dominant land use in the region and the primary reason for the presence of large, intact areas of native mixed-grass prairie (Kennedy and McMaster, 2003). Dry-land farming is also important in the local economy, and tourism is starting to play a larger role in the wake of widespread depopulation, the establishment of Grasslands National Park in the east, and important fossil discoveries, including Scotty the *Tyrannosaurus rex* in the west.

The Frenchman area was chosen as the study site because it met many biological, social, and climatic criteria. Compared to other watersheds on the Canadian Prairies, the Frenchman watershed is more biologically diverse. The river system is a productive waterway that passes through a wide range of landscapes (Plate 1) and drains southward into the Missouri, providing habitat for species that are more typical of southern locations (Kennedy and McMaster, 2003). The Frenchman region also met important social criteria, including high cultural diversity, high diversity of land use, and good potential for local involvement. A final, overarching reason for studying the Frenchman watershed is the probability of local climatic conditions becoming warmer and drier because of global climate change (Sauchyn et al., this issue). The effects of warmer, drier conditions on local water use are likely to be very significant because the entire flow in the Frenchman River is already allocated for various purposes and uses, including commitments to the United States.

The FRBP is modeled after other watershed studies, especially the Rideau River Biodiversity Project (RRBP) in eastern Ontario (Johnson et al., 2003), which ran from 1998 to 2000 and led to the development of the Rideau River Roundtable (http://www.nature.ca/rideau). The RRBP was initially designed to assess the biodiversity of the Rideau River by collecting information about water quality and eight different biotic groups, namely phytoplankton, macrophytes, molluscs, general invertebrates, amphibians, reptiles, fish and aquatic birds. The second primary objective was to reconcile local needs with the long-term sustainable management of its biodiversity. The Rideau model (Johnson et al., 2003) was a useful reference during the development of the FRBP because of its emphasis on the need to involve local stake-holders, but the wide range of taxonomic groups studied through the RRBP was problematic. To avoid local concerns about vertebrate species at risk, especially fish, the Steering Committee decided to restrict the focus of the FRBP to aquatic invertebrates, such as insects and mollusks, with an emphasis on identifying and using certain indicator species or groups of species to monitor changes in the ecosystem. Plans for the social research associated with the FRBP were influenced by the "citizen-centered" approach that was used in the San Fancisquito Creek project (Karl and Turner, 2002). The final FRBP model (Figure 1) assumes that information about aquatic biodiversity can shed light on the health of a river system and on social values and norms that affect local management decisions.

The FRBP model (Figure 1) reflects a central tenet of sustainable development, where "a healthy society gives equal attention to ecological sustainability, economic development, and social justice because they are all mutually reinforcing" (Marten,

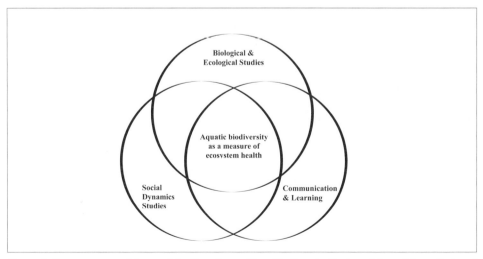

Figure 1. Components and focus of the Frenchman River Biodiversity Project. Adapted from Karl and Turner (2002).

2001: 9). Adopting this model affected many aspects of the project, including the central research question. An early version of the question, which asked whether the Frenchman River is healthy and being used sustainably, could be addressed through regular environmental monitoring, with little or no community involvement. As the project evolved, the question became "How can we keep the Frenchman River healthy?": a question that is more in line with CBEM principles. The current question makes it clear that local residents are playing an active role in the project and gives more weight to social attitudes and interactions that might affect the health of the watershed.

Gaining the Trust of Local Communities

Once the scope and focus of the research had been defined, the stage was set for community meetings. Two initial meetings were held, one in Eastend and one in Val Marie, to determine the feasibility of the study and to garner local support. These events, which were part information session and part roundtable discussion, were open to the public and held in the evening outside of calving season. Both meetings were widely advertised through rural municipality offices and other outlets and both attracted about 30 people (all adults). The discussions that ensued were challenging because of skepticism about the value of the work and fears about federal legislation designed to protect species at risk, coupled with a general mistrust of outside agencies (Diaz and Nelson, this volume). Specific concerns voiced by the participants, who ranged from homemakers and teachers to business owners and ranchers, included: negative public perception around the impact of cattle ranching;water allocations to the US (the Frenchman is an international watershed); low flows and the need to fence cattle off waterways; risks to lifestyle and the ranching industry, especially around access to landowner information and potential impacts of the Canadian Species At Risk Act (SARA); uncertainty about on-going institutional and financial support; time commitments on the Steering Committee; the source of existing financial support; and whether more information would be an advantage in dealing with outside regulators.

The interim Steering Committee took three steps to address these concerns, especially the high levels of skepticism and concerns about land-owner confidentiality. First, a data management agreement was proposed based on a model used for research on endangered species in southern Alberta (R.A. Sissons, unpublished data). Under this agreement, plot locations would be "fuzzed" on maps and not disclosed either in writing or in person to ensure landowner confidentiality. In return, researchers would be allowed to send precise location information to the Saskatchewan Conservation Data Centre (SK-CDC) for archival purposes.

Second, information was provided about provincial policies for the management of biological data through the SK-CDC. These policies stipulate that data must be withheld where releasing it could cause harm (financial or otherwise) to individuals, agencies or the resource; no site-specific data are to be released for endangered species; precise land locations in the database must be "degraded" (presented with a reduced resolution) when records are released; project participants can have full access to the data, but are bound by the above restrictions and can only publish the results of analysis (as opposed to specific land locations); and personal information does not have to be recorded, and if it is, it can only be released with consent (Kevin Murphy, pers. comm.).

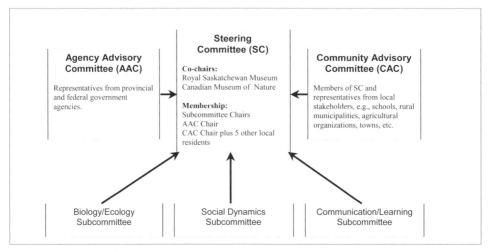

Figure 2. Committee structure of the Frenchman River Biodiversity Project.

Third, a revised committee structure was developed that would give local residents a measure of control over the project. This structure (Figure 2) includes a Steering Committee that is at least 50% local and supported by a pair of advisory committees that represent relevant agencies on one hand and local communities on the other.

This set the stage for a second round of public meetings, which were similar to the first ones in their organization but attracted fewer people (see description below). Towards the end of these meetings, the community members in attendance were polled by secret ballot to determine whether they thought the project should proceed or not. There was some risk in this, but it seemed to be the most effective way to assess the thinking of those in attendance. All local participants voted, no ballots were spoiled, and the votes were 88% in favour of the project (21 of 24 ballots) in one community and 67% in favour (12 of 18 ballots) in another. The votes generated some resentment, since they revealed that vocal people who opposed the project were in the minority and the results reflect the thinking of only a small portion of each community. Nevertheless, local residents appeared to gain an increased sense of ownership over the project. During the first meeting of the new Steering Committee, which took place several months later, local representatives began to refer to the work as *our* project rather than *your* project.

Researcher Concerns about Confidentiality

Discussions about the need to protect land-owner information left project researchers concerned about the impacts that a restrictive data management agreement might have on their ability to publish their findings. To assess the level of risk, we asked the editors of seven leading, peer-reviewed journals in aquatic and environmental science about the level of information they would require in order for a paper to be published. Specifically, we asked the editors to indicate which of the following scenarios their journal would accept, or, to suggest other options, if necessary:

• A statement similar to "a prairie river in southern Canada."

- Naming the river system but not delineating where samples or plots were taken or located.
- Naming the river with locations plotted but obscured to the point where no one could relocate the plots by the map.

There was no intention to publish the editors' responses when this request was made, so we are not in a position to reveal the names of specific editors or journals, but we can talk about the results of the survey in general terms. One editor was unable to answer because the topic of landowner confidentiality was being discussed by the professional organization associated with the journal in question. Most of the others (5 of 6) indicated that some level of obscurity in the plot location data (option #2, or some combination of #2 and #3) would be acceptable as long as an explanation was provided. Several noted that naming the river and giving approximate locations would suffice and that a map would only be essential if it helped readers understand some aspect of the study, such as the experimental design. Others offered specific suggestions, such as providing longitude and latitude without identifying exact locations and using a footnote to let readers know that the editors and manuscript referees had been provided with location details during the review process (a minimum requirement in one case). One editor was especially sympathetic to the issue, noting that "it is reasonable to try to respect [the] wishes [of the landowners] and protect [them] in return for their cooperation... I think this journal could fudge a bit on locations ... and I would hope that most journals would do so."

Only one editor indicated that there would be a problem, noting the importance of "sharing information with other scientists working in the same area" and the possibility that researchers might want to relocate sites in the future to assess changes. Even here, however, the editor expressed empathy for landowner concerns, and encouraged the submission of papers that might be useful in developing relevant policies.

The brief responses we received from these editors should not be viewed as widespread support for the publication of research findings that provide limited or "fuzzed" information about sampling locations. More stringent policies or opinions may be norm for journals in other fields, e.g., terrestrial ecology, and differences may develop over time as the editors we contacted are replaced. As a current snapshot, however, the responses suggest that aquatic biologists should feel comfortable with data management protocols that restrict the publication of landowner information.

Fostering Trust Within Communities

The success of a CBEM project can hinge on the degree to which communities trust the local people that offer, or are selected, to represent and speak for them. In other words, local politics can be a major factor. Outside agencies can do little, if anything, to increase inherent levels of trust, but this may not be problematic on the Canadian Prairies. People in rural prairie communities tend to have high levels of trust for local residents, unless they are the subjects of pre-existing prejudices (Diaz and Nelson, this issue). The main challenge for outside agencies involves ensuring that the outcomes of the project are in line with real community needs, encouraging local participants to speak on behalf of the project, and supporting them when opportunities to speak arise.

Three situations in the development of the FRBP illustrate how this challenge can be addressed. First, during the initial round of community meetings, project proponents suggested that local communities might benefit from having a snap-shot of the impacts of current human activities and, eventually, a community-based mechanism for addressing issues that may affect the river system now and in the future. Ultimately, neither of these arguments carried much weight. The turning point occurred when local participants suggested that their communities would be able to take a critical look at current practices and their impacts rather than wait-ing for an outside agency to lead a study and form their own conclusions or to impose recommendations based on studies done elsewhere. In other words, local residents appeared to recognize the value of the work once a community need had been articulated by other local people. This may reflect the strong sense of auton-omy and pride associated with the ranching culture of southwest Saskatchewan. It also highlights the importance of giving local residents time to reflect on and artic-ulate why a project might be valuable for them, and to adjust the direction of the project based on their needs.

The other two situations are mostly about fostering trust through effective extension activities. When a local member of the Steering Committee spoke about the project at a related community meeting, he faced a number of pointed questions from landowners who assumed that project researchers had ventured onto private land without permission. Through a series of emails that started the next day, the researchers and project coordinator provided him with details about the sampling program and the landowner contacts, which made it clear that FRBP researchers had not been involved and had in fact been very conscientious about securing landowner permission. Finally, local participants have been encouraged to speak on behalf of the FRBP by providing them with information about confer-ences they might attend, such as a provincial conference on watershed manage-ment issues, and by creating a space called "Local Voices" in the project newsletter.

Concluding Remarks

This article examines many of the challenges involved in fostering mutual trust as a "top-down" idea is transformed into a community-based, "bottom-up" study of environmental and social sustainability. Insights gained through the development of the FRBP suggest that local communities are more likely to commit to a project that originated elsewhere when: the project outcomes are clear, practical, and have obvious long-term benefits to the community; the opinions and concerns expressed by local residents are respected and recorded; tangible steps are taken to address local concerns, such as the development of a data collection protocol that protects landowner confidentiality; and local stakeholders have a measure of control over the direction of the project.

The other steps we took to foster mutual trust around and within the FRBP sug-gest that aquatic biologists should be able to publish research papers that provide limited information about sampling locations, given the current thinking of journal editors in their field. Our efforts also imply that democratic processes can be effec-tive in transferring ownership of the project to local stakeholders, but they may also have a polarizing effect if vocal people are expressing minority opinions.

Projects like the FRBP are important because they address water-related issues, a persistent concern on the northern prairies. Water issues are likely to receive even more attention in the near future with the advent of warmer, drier conditions due to climate change (Sauchyn et al., this issue). Initiatives like the FRBP and other CBEM projects could help to address these issues and foster the development of sustainable communities, especially if they catalyze the formation of local stewardship groups that contribute to monitoring, lobbying, and other critical activities. Our efforts to develop the FRBP have shown that fostering trust at many levels is a critical and challenging part of the process. The local culture needs to be considered, and levels of project support across and within target communities need to be assessed by using suitable survey tools and methodologies. The FRBP is still unfolding, and is currently involved in school-based monitoring programs and additional research planning. Thus, we expect to learn even more about circumstances and activities that can foster or compromise community trust. Our hope is that the steps we have taken and the insights we have gained will be helpful to other organizations interested in biodiversity conservation, community involvement, and policies that affect watershed management.

Acknowledgements

This paper was greatly improved by comments from Todd Radenbaugh and Stan Morse, and by input from André Martel, who took part in the initial planning discussions for the FRBP. Research and planning for the FRBP has been supported by funding from the EJLB Foundation, Environment Canada, the Salamander Foundation, the Saskatchewan Heritage Foundation, the Natural Sciences and Engineering Research Council, and the Social Sciences and Humanities Research Council.

References

Ack, Bradley L., Carol Daly, Yvonne Everett, Juan Mendoza, Mary Mitsos and Ron Ochs. 2001. "The Practice of Stewardship: Caring for Healing Ecosystems and Communities," *The Journal of Sustainable Forestry* 12, nos. 3–4: 117–141.

Costanza, R., R. d'Arge, R. de Groot, S. Farber, M. Grasso, B. Hannon, K. Limburg, S. Naeem, R.V. O'Neill, J. Paruelo, R.G. Raskin, P. Sutton and M. van de Belt. 1997. "The Value of the World's Ecosystem Services and Natural Capital," *Nature* 387: 253–260.

Dietz, T. 2003. "What is a Good Decision? Criteria for Environmental Decision Making," *Human Ecology Review* 10: 33–39.

Engel, S.R. and J.R. Voshell Jr. 2002. "Volunteer Biological Monitoring: Can It Accurately Assess the Ecological Condition of Streams?" *Amerian Entomologist* 48, no. 3: 164–177.

Flora, Cornelia. 1995. "Social Capital and Sustainability," *Research in Rural Sociology and Development* 6: 227–246.

Gray, G., M.J. Enzer and J. Kusel (eds.). 2001. *Understanding Community-based Forest Ecosystem Management.* New York, London and Oxford: Food Products Press (an imprint of The Haworth Press, Inc).

Johnson, M.C., M. Poulin and M. Graham. 2003. "Towards an Integrated Approach to the Conservation and Sustainable Use of Biodiversity: Lessons Learned from the Rideau River Biodiversity Project," *Human Ecology Review* 10: 40–55.

Karl, H. A. and C. Turner. 2002. "A Model Project for Exploring the Role of Sustainability Science in a Citizen-centered, Collaborative Decision-making Process," *Human Ecology Forum* 9: 67–71.

Martel, A.L., D.A. Pathy, J.B. Madill, C.B. Renaud, S.L. Dean and S.J. Kerr. 2001. "Decline and Regional Extirpation of Freshwater Mussels (*Unionidae*) in a Small River System Invaded by *Dreissena polymorpha*: The Rideau River, 1993–2000." *Canadian Journal of Zoology* 79: 2181–2191.

Marten, G.G. 2001. *Human Ecology: Basic Concepts for Sustainable Development.* Sterling VA: Earthscan Publications Ltd.

Kennedy, R. and G. McMaster. 2003. *Overview of the Frenchman River Watershed: A Report of the Frenchman River Biodiversity Project.* Regina and Ottawa: Saskatchewan Culture Youth and Recreation, and Canadian Museum of Nature.

Moote, M.A., B.A. Brown, E. Kinsly, S.X. Lee, S. Marshall, D.E. Voth and G.B. Walker. 2001. "Process: Redefining Relationships," *Journal of Sustainable Forestry* 12, nos. 3–4: 97–116.

Putnam, R. 2001. "Social Capital: Measurement and Consequences," *Isuma* 2, no. 1: 41–52.

Rapport, D. 2000. "Is It All About US?" *Ecosystem Health* 6: 169–170.

Ray, Christopher. 1999. "Towards a Meta-framework of Endogenous Development: Repertoires, Paths, Democracy and Rights," *Sociologia Ruralis* 39, no. 4.

Schindler, B. and K. Aldred Cheek. 1999. "Integrating Citizens in Adaptive Management: A Propositional Analysis," *Conservation Ecology* 3, no. 1:
online URL: http://www.consecol.org/vol3/iss1/art9

Webler, T. and S. Tuler. 2001. "Public Participation in Watershed Management Planning: Views on Process from People in the Field," *Human Ecology Forum* 8: 29–39.

Appendix 1

Needs identified by the interim Steering Committee for the Frenchman River Biodiversity Project (FRBP):

• Researchers and land-owners need to know about aquatic biodiversity now so they can detect changes in the future, including the effects of climate change.

• Prairie watersheds and communities are affected by many factors, including land- and water-use decisions made by individual farmers and ranchers. Local people need to be involved in studies of ecosystem health because they are familiar with issues that affect their watershed, and they have a vested interest in how problems are addressed.

• Many lines of communication have to be established and maintained so all of the stakeholders associated with a watershed have a chance to assess and compare the effects of different management options.

Guiding principles of the FRBP:

• By taking a watershed perspective, project participants recognize that land use, soil, and water are interrelated and that events and activities upstream affect everything downstream.

• Comparisons of water quality and aquatic biodiversity between different locations need to be qualified, since the effects of land use are likely to vary within a watershed due to the effects of vegetation and soil condition.

• People are an integral part of their ecosystems, contributing to biodiversity in the broadest sense. Introduced species also contribute to the biodiversity of a watershed and can alter the way ecosystems function.

• To be sustainable, human activities have to meet current needs without compromising the welfare of future generations. Sustainability involves reflecting on decisions made in the past and recognizing that social and economic activities are constrained by ecological limits.

CHAPTER 7

Re-evaluation of Native Plant Species for Seeding and Grazing by Livestock on the Semiarid Prairie of Western Canada

P.G. Jefferson, A.D. Iwaasa, M.P. Schellenberg and J.G. McLeod

Introduction

A Brief History

European settlement of the Mixed Grassland ecoregion of southwestern Saskatchewan and southeastern Alberta occurred in two waves. Corporate ranching based on very large tracts of native grasslands leased from the federal government followed the construction of the Canadian Pacific Railway in the 1880s, but inadequate cattle feed during extreme winters resulted in large losses of livestock due to death and subsequent bankruptcies (McGowan, 1975). Despite concerns about low and variable precipitation and low organic-matter soils that are marginal for annual crop production, the federal government's settlement policy was changed by 1900 to encourage the development of small farms and a supporting rural infrastructure such as railway branch lines. Immigrant farmers, or "sodbusters," received 160 acres (64 hectares) of land in exchange for plowing native grass for crop production (mainly wheat, *Triticum aestivum* L.) for export to Europe. This culminated in the plowing of 80% of the native grasslands of this ecoregion in Saskatchewan and Alberta (Selby and Santry, 1996). When the unsustainable nature of some homesteads became evident during the major droughts of the 1930s, introduced or exotic forage species, such as crested wheatgrass (*Agropyron cristatum* L. *Gaertn.* and *A. desertorum* (Fisch. Ex Link J.A. Schultes), were used for reseeding millions of hectares (Gray, 1996).

Introduced forage grass species from central Asia have been extensively studied since the 1950s for improved adaptation, forage yield, forage quality, and persistence under ruminant grazing pressure. At the same time, in situ native species were considered to have low forage yield, poor seed production, low forage quality, and low carrying capacity for grazing by domestic livestock. Over the last decade, however, there has been a renewed and growing interest in native plant species, contributing to the formation of the provincial Prairie Conservation Action Plans (PCAP Partnership, 2003), the popular Native Plant Summit conferences, and specialist groups such as the Native Plant Society of Saskatchewan. This revival can be attributed to several emerging trends and changes in the agriculture industry, which include an increasing demand for wildlife habitat; an increasingly ecological perspective on grassland management by cattle ranchers; a greater respect for the role of grasses in soil organic carbon sequestration; and a better understanding of the invasive characteristics of some introduced forage species.

Context and Purpose

Native prairie has higher floral biodiversity than modern, monoculture crop-ping systems, but conserving this biodiversity on ruminant livestock pastures presents many challenges. Little biodiversity data are available to compare undisturbed prairie ecosystems to agro-ecosystems (Paoletti et al., 1992). On a continental scale, there has been a 64% reduction of the Mixed-Grass prairie. In Alberta, Saskatchewan, and Manitoba, native prairie grasslands have declined by 61%, 81%, and 99%, respectively. The Tall-Grass prairie in Manitoba has been almost completely converted to crop production, and Saskatchewan has lost all but 6% of its Mixed-Grass prairie (Samson et al., 1998). Previous efforts to revegetate these areas have been hampered by the invasion of exotic species in native rangelands.

Concern about native grassland has also created opportunities. Many native plants were used by the first peoples of Canada for medicinal purposes and there is renewed interest in such plants as a source of pharmaceutical and nutritional products (Lewis and Elwin-Lewis, 2003). A recent market assessment of native plants in Saskatchewan commissioned by the Native Plant Society of Saskatchewan (Solutions 2000[+], 1997) forecasts a 15% yearly increase in market size for native plants including species with medicinal value. In urban settings, there has also been a growing demand for native plants for low-input xeriscaping.

The introduction of the Species at Risk Act 2003 in the federal parliament, an act to promote the preservation of endangered species in Canada, has heightened awareness of the need for the preservation of native mixed prairie rangelands and habitats. At the same time, private and public organizations that manage native prairie landscapes have recognized the importance of revegetating disturbed sites with native rather than introduced plant species. For example, Ducks Unlimited Canada has promoted the value of seeded native species for ground-nesting water-fowl habitat on land adjacent to permanent water bodies. In the ecologically sensitive Great Sandhills of southwestern Saskatchewan, the petroleum industry is now revegetating pipeline rights-of-way, well sites, and other disturbed areas with native species. In Alberta, the use of native species has been increasing steadily for public and private land reclamation, resulting in the need for new guidelines (Native Plant Working Group, 2001).

Lessons about seeding with native plants can be learned from the Conservation Reserve Program (CRP) in the United States, which promotes the revegetation of land that is marginal for annual crop production and emphasizes native species. One benefit of the CRP is increased soil organic carbon concentration on fragile soils in the northern Great Plains (Gebhart et al., 1994). A forage seeding program for Canada has been funded and will support conversion of marginal crop land to perennial vegetative cover, but it also has a component for re-seeding native species (Agriculture and Agri-Food Canada, 2003). Over time, soil reseeded with native species could help Canada meet its international commitment to reduce greenhouse gas emissions (Environment Canada, 2002) by removing carbon from the atmosphere via photosynthesis, and storing it as soil organic matter carbon (Christian and Wilson, 1999; Janzen et al., 2000).

Cattle producers' appreciation for native rangelands has increased, reflecting years of range experience and training in intensive grazing management based on ecological principles. Native rangelands in the Mixed Grassland ecoregion can support good summer weight gains on yearling steers (Karn and Lorenz, 1983) and are similar to some introduced grasses (Smoliak and Slen, 1974; Hofmann et al., 1993). Given that stocking rate, grazing rotation among paddocks, and grazing duration can be managed to improve range condition and maximize economic productivity (Hart et al., 1988), seeded native species are becoming important forage sources that complement existing native rangelands for improved grazing management and range condition. Moreover, as changes to transportation and input costs make grain production uneconomical on some soils, the investment in re-seeding native species for summer, fall, and winter grazing by cattle has grown. Fall and winter grazing of beef cattle, particularly pregnant non-lactating beef cows, is one strategy to reduce the cost of beef production in this region.

The purpose of this paper is to review revegetation research involving native plant species of the semiarid prairie region in light of these trends. A second objective is to identify opportunities for further research and development of native plants for seeding on the prairie landscape.

Attributes of Native Versus Introduced Forage

Adaptation

Establishment of plant species from seed is key to the revegetation of disturbed sites, development of new pastures for livestock grazing, or range improvement. Previous research suggests that native grasses are difficult to establish. Kilcher and Looman (1983) report very low establishment of big bluestem (*Andropogon gerardii* Vitman.) and prairie sandreed (*Calamovilfa longifolia* (Hook.) Scribn.) at Swift Current, Saskatchewan as represented in Table 1 (lat. 50°16' N, long. 107° 44' W, 825 m elevation). More recent results (Jefferson et al., 2002; Jefferson, unpublished) indicate that varieties of these native grasses from Montana and North Dakota can be successfully established from seed at Swift Current (Table 1) and other sites in western Canada (Jefferson et al., 2002), although cool-season (C3) grasses established better stands than warm-season (C4) grasses. Jefferson et al. (2002) also observe good seedling vigor and excellent stand establishment for other warm-season species (Table 1). The earlier work of Kilcher and Looman (1983) was based on varieties from Kansas, Nebraska and Colorado, whereas Jefferson et al. (2002) studied varieties from Montana and North Dakota. These states are adjacent to southern Saskatchewan and can provide varieties that are better adapted to western Canada than varieties from further south (Tober and Chamrad, 1992).

Lawrence and Ratzlaff (1989) conclude that native grass species were not persistent when seeded at Swift Current, Saskatchewan. The native grasses they tested included five slender wheatgrass (*Elymus trachycaulus* [Link Gould ex Shinners]) varieties, four awned wheatgrass (*Agropyron subsecundum* [Link Hitchc.]) varieties, two northern wheatgrass (*Elymus lanceolatus* [Scribn. & J.G. Sm.]) varieties, and one streambank wheatgrass (*Elymus lanceolatus* [Scribn. & J.G. Sm.] *Gould* ssp. *lanceolatus*). They compared these species to three introduced forages, namely crested wheatgrass, intermediate wheatgrass (*Elytrigia intermedia* [Host] Nevski), and meadow bromegrass (*Bromus riparius* Rehmann), in terms of forage yield and plant density. They reasoned that slender wheatgrass lacks winter hardiness or persistence as it died in the fifth year after establishment. This conclusion was then extrapolated to all native species relative to introduced grasses

Table 1. Establishment of Native Grasses in Three Studies.

Common name	Latin name*	Kilcher and Looman, 1983	Jefferson et al., 2002	Jefferson, unpublished data
Introduced species			%	
Crested wheatgrass	Agropyron desertorum (Fisch. ex Link) J.A. Schultes	100	–	100
Russian wildrye	Psathyrostachys juncea (Fisch.) Nevski	100	–	100
Native species				
Northern wheatgrass	Elymus lanceolatus (Scribn. & J.G. Sm.)	90	98	–
Western wheatgrass	Pascopyrum smithii (Rydb.) A. Love	80	94	–
Green needlegrass	Nasella viridula (Trin.) Barkworth	90	98	96
Big bluestem	Andropogon gerardii Vitman.	5	89	94
Little bluestem	Schizachyrium scoparium (Michx.) Nash	3	44	–
Prairie sandreed	Calamovilfa longifolia (Hook.) Scribn.	5	68	95
Indiangrass	Sorghastrum nutans (L.) Nash	11	32	80
Switchgrass	Panicum virgatum L.	–	86	100
Sand dropseed	Sporobolus cryptandrus (Torr.) Gray	6	–	23
LSD 0.05 18		NR **	7	21

* Nomenclature after Alderson and Sharp 1994: 19
** NR - Not reported: 20

for the semi-arid region of western Canada. However, these authors failed to recognize that slender wheatgrass and awned wheatgrass are short-lived perennials that die out in three to five years. Thus their primary conclusion, and its extrapolation to all native species, was erroneous. The authors also failed to acknowledge that northern wheatgrass exhibited no stand loss in this trial (Lawrence and Ratzlaff, 1989).

Current research projects (Waller et al., 1994; Schellenberg and Jefferson, 1998; Banerjee and Schellenberg, 2000) have succeeded in establishing various native grass, shrub and forb ecotypes in southern Saskatchewan. More research is needed to establish proper seeding and establishment procedures and optimum seeding density; determine the advantages of mixtures versus species monocultures; determine the role that grazing animals play in perpetuating local plant populations; develop proper grazing management strategies; and evaluate the productivity of the stands in the short and long term.

Forage Production

Direct comparisons of native and introduced species at Swift Current suggest that introduced species produce more above-ground forage biomass than native species (Lawrence and Troelsen, 1964; Lawrence, 1978; Kilcher and Looman, 1983; Lawrence and Ratzlaff, 1989). However, 90% of the variation in forage

Table 2. Forage yield (kg ha^{-1}) of introduced and native forage grass species clipped three times per season for 1991 and 1992 and once per season for 1993 and 1994 at Swift Current, Saskatchewan. Least significant difference (LSD) values and probability of contrasts between groups of species are shown.

Species	1991 May 30	1991 Jul 9	1991 Aug 20	1992 May 28	1992 Jul 15	1992 Oct 22	1993 Jul 16	1994 Jul 14
Introduced grasses								
Crested wheatgrass	3810	2800	700	840	560	500	920	4430
Intermediate wheatgrass	2500	3370	2050	270	1670	100	960	4050
Russian wildrye	1740	1950	850	960	710	270	1220	3910
Tall Fescue	1740	4240	1470	720	1250	650	760	4210
Tall Wheatgrass	2030	4690	1150	700	1530	610	1070	4540
Native cool-season								
Beardless wildrye	400	1080	410	50	670	160	460	4130
Green needlegrass	650	1730	1160	460	920	410	1120	5400
Junegrass	280	20	–	240	0	–	150	3300
Native warm-season								
Big bluestem	–	200	220	–	660	100	410	4830
Blue grama	240	280	280	–	800	430	400	4690
Indian ricegrass	–	–	–	–	70	60	230	3110
Indian grass	–	230	780	–	590	280	–	3590
Prairie sandreed	190	560	270	–	600	180	660	5400
Side-oats grama	–	190	580	–	260	300	–	3150
Switchgrass	–	560	460	–	1340	310	650	4900
Sand dropseed	–	–	–	–	–	–	500	3200
LSD	720	900	920	600	560	350	NS	1200
P>F Introduced vs Native	<0.001	<0.001	0.001	0.007	<0.001	0.024	–	0.75
P>F cool- vs warm-season	0.487	0.013	0.228	N.A.	0.519	0.669	–	0.565

production among native and introduced species in Kilcher and Looman (1983) is explained by variation in stand establishment. Similarly, variation in forage yield between native and introduced species can be attributed to variation in stand establishment at a site in the intermountain region of the US (Asay et al., 2001). Other studies (Coupland, 1974; Dubbs et al., 1974; Hanson et al., 1976) show native grasses comparing favorably with introduced species. Knowles (1987) reports that western wheatgrass excelled in pasture yields and deserves attention for seeding long-term pastures. He also notes that introduced grasses were less productive during the summer period than native species, which continue to grow. Jefferson et al. (1999) report that sites seeded with an introduced species mixture produced more forage than sites seeded with native species in Manitoba and Alberta, while there was no difference between the two types in Saskatchewan.

Direct comparisons of native cool- and warm-season grasses with five introduced cool-season grasses indicate that yield differences are affected by the timing and frequency of forage harvest (Jefferson, unpublished data). Introduced species produced superior forage yield when harvested in late May in 1991 and 1992 compared to cool-season native grasses at the first of three harvests per year (Table 2). The native warm-season grasses had not produced measurable forage biomass by the end of May each year. By 1994, one year after the harvest timing had been changed to once per growing season (mid-July), native grasses had produced forage yields equivalent to the introduced species (Table 2). These results suggest that previously reported forage yield comparisons may have been biased by harvests that occurred too early and too frequently during the growing season.

The biomass productivity advantage that introduced species have over native species appears to depend on soil fertility (Johnston et al., 1968). Introduced species yield more forage than native range only when fertilizer is applied; unfertilized introduced species actually yield less than unfertilized native range (Johnston et al., 1968; Knowles, 1987). This has implications for ranchers in the semiarid prairie region, where the application of forage fertilizer is economically risky due to variation in rainfall amount and inconsistent responses to fertilizer application. Earlier forage biomass productivity comparisons may reflect the fact that the species were grown on the Swift Current Research Centre, where available soil nutrients and weed control are significantly above average. Also, since the yield of introduced grasses peaks at two to three years after seeding and declines thereafter (Knowles, 1987), earlier forage yield comparisons may have been biased by short (one to three years) trials. Introduced species likely have less of a biomass productivity advantage over native grasses on semiarid prairie pastures than that reported from research centre trials. Crested wheatgrass may be an exception, as it appears to maintain its above-ground productivity advantage over native rangeland for long periods without fertilization (Smoliak et al., 1967). Crested wheatgrass pastures can be two to ten times more productive than adjacent native range (Smoliak et al., 1967; Looman and Heinrichs, 1973).

In summary, forage yield production comparisons of native and introduced species are difficult to interpret due to the interaction of soil type, fertility, testing period, harvest management (date), climate, and possibly other edaphic factors. We conclude that the advantage in forage productivity of introduced forage species under intensive agronomic management is the primary reason for continued use of these species for hay and spring pasture production in the semiarid prairie region.

Grazing and Forage Quality

Native grasses evolved to withstand the extreme environmental conditions and buffalo grazing pressures of the original mixed grass prairie, while introduced grasses evolved in other parts of the globe. Kamstra (1973) reports that individual native grasses each have specific seasonal growth characteristics that can vary their nutritional quality as phenological development proceeds. Therefore, the usual definitions do not consistently apply to range grasses, and direct comparison of quality components between introduced and native grasses at equivalent maturity stages cannot be made. For example, many range grasses produce seed only during favorable years. In addition, Knowles (1987) and Coupland (1974) reported that western wheatgrass and northern wheatgrass exhibit a greater degree of senescence compared with cultivated grasses. Although the nutritional contents and digestibility can vary among different native species during the growing season (Abouguendia, 1998), cattle producers have recognized that several native species common to the prairies of western Canada retain a relatively high nutritive value during late summer and throughout fall and early winter. In addition, these native species are able to preserve their physical form; their stems and particularly leaves

Table 3. Native grass and shrub species that exist in the prairie areas of Saskatchewan and Alberta that demonstrate the ability to cure to a greater or lesser degree (Pigden, 1953).

Common Name(s)	Latin Name
Speargrass, Needle-and-thread	*Stipa comata* Trin. & Rupr.**
Blue grama grass	*Bouteloua gracilis* (Willd. ex Kunth), Lag. ex Griffiths*
Western wheatgrass	*Pascopyrum smithii**
Northern wheatgrass	*Elymus lanceolatus**
June grass	*Koeleria gracilis* Pers.**
Rough fescue	*Festuca hallii* (Vasey) Piper**
Porcupine grass	*Stipa spartea* Trin.**
Salt sage	*Atriplex nuttallii* S. Wats.**
Winterfat	*Krascheninnokovia lanata* (syn. Eurotia *lanata* (Pursh) Moq.)**

* Nomenclature after Alderson and Sharp, 1994
**Nomcmclature after Budd et al., 1987

do not decompose to any extent for some months after growth has ceased (Table 3). Referred to by cattle producers as "curing," this remarkable property is one reason native pastures can be grazed by cattle later in the grazing season or in moderate winter conditions with little or no supplemental feeding. Curing normally occurs during late July but can also take place in mid-June or late August, depending on the season (Pigden, 1953). In economic terms, the excellent curing ability of many native species can reduce animal feed costs by shortening the winter feeding period.

Cattle ranchers may not have a sufficient land base of native range to exploit the summer and fall grazing potential of native species. Some are interested in seeding native species mixtures specifically for late-season grazing. Based on seeded native-pasture studies in 1993 and 1994 at Swift Current, calf weight gains on pasture in fall and early winter were nearly as good as those observed in feedlots (Jefferson et al., 1997). Estimates of grazing capacity (grazing days per ha) on the native grasses were similar to those reported for introduced grasses. Of the native grasses, a monoculture of western wheatgrass exhibited better forage quality for winter grazing than a mixture of northern wheatgrass, western wheatgrass and green needle grass. In another study, the digestibility of western wheatgrass harvested in September was 14% higher than that of northern wheatgrass (Jefferson et al., 2004). Western wheatgrass maintained good forage quality throughout the grazing season in North Dakota (Frank and Bauer, 1991; Hofmann et al., 1993) but can have low quality when harvested at a late phenological stage (Smoliak and Bezeau, 1967). However, other results indicate that western wheatgrass has better forage quality than northern wheatgrass and has good potential for fall grazing (Jefferson et al., 2004; Knowles, 1987).

The quantity and quality of forage produced on native rangelands are highly cyclical within and between years, and one would expect similar variation for seeded native mixtures. Precipitation, plant species, and the proportion of cool- and warm-season species would affect overall forage quality of seeded native pastures at any given time and as a result lead to seasonal patterns of livestock gains. In general, diets from dormant native grasses contain 4% to 7% crude protein, with higher concentrations occurring from late summer to early fall, and lower concentrations occurring from late fall through winter. Plants in a vegetative state, and some shrubs such as winterfat can contain over 10% crude protein (Abouguendia, 1998) in late fall or early winter.

Native rangeland consists of a diverse community of forbs and shrubs, so the forage quality of seeded native species mixtures may be enhanced by their inclusion. Shrub species, such as winterfat (*Krascheninnikovia lanata* [Pursh] Meeuse & Smit syn. Eurotia lanata [Pursh]) Moq.), have superior forage quality for late fall and winter grazing (Smoliak and Bezeau, 1967; Abouguendia, 1998). Winterfat could be successfully established from seedings (Romo et al., 1997; Schellenberg, unpublished data) and could contribute to significant improvements in cattle gains on fall and winter pastures of seeded native species. Gardner's saltbush (*Atriplex gardneri* [Moq.] D. Dietr.) has similar nutritional qualities to winterfat (Smoliak and Bezeau, 1967) and could also contribute to cattle nutrition for fall grazing.

Livestock productivity during the grazing season can be predicted from forage quality estimates such as digestibility and protein. The digestibility of most native grasses from May until September (Abouguendia, 1998) is able to meet the nutritional requirements of a lactating beef cow or a growing yearling (0.45 kg gain d^{-1}) based on the guidelines developed by Holechek and Herbel (1986). Phosphorus (P) is the most limiting mineral to range livestock production, and adequate plant P concentrations for growth or lactation occurs only for a brief period early in the growing season, for example, during the vegetative phase of plant growth. Calcium concentrations of native grasses are adequate for maintenance, growth, and lactation throughout the year (Abouguendia, 1998). Cattle grazing on only native grasses in late fall, winter, and early spring may require additional protein and P supplementation. Phosphorus supplementation may also be needed during the growing season to satisfy the needs of lactating cows and young cattle (Abouguendia, 1998). However, by including forbs and shrubs in cattle diets, cattle performance can be improved during periods when grasses are dormant and low in quality (Holechek et al., 1989). Several studies reviewed by Holechek et al. (1989) find that leaves from forbs and shrubs contain more protein (P), and cell solubles than do grasses at comparable stages of maturity. Schellenberg and Jefferson (1998) report that the native woody shrubs, winterfat and Nuttall's saltbush (*Atriplex nuttallii* S. Wats.), which are found in southwest Saskatchewan and southeast Alberta, retain their nutritive value well into the fall and early winter (protein concentration was 11.5%). Although information is limited on associative effects between forage species on intake, Milchunas et al. (1978) report that shrub species in the diet may increase the digestibility of grasses and increase the overall digestibility of the total diet. During winter, shrubs with a higher protein concentration could improve the intake of grasses with protein levels below 7% by providing rumen microbes with a source of nitrogen (Cordova and Wallace, 1975). Arthun et al. (1992) concludes that adding forbs or shrubs with low quality grass had a similar effect on the ruminal digestion kinetics and fermentation process of cattle as including alfalfa hay.

Research by Abouguendia (1998) reveals wide variations in nutrient contents and digestibility both among growth forms and among species in these forms. Therefore, it is important to identify the dominant species and their proportions in each field in order to make efficient use of the available nutrient supply over the entire grazing season and across the landscape. In southern Saskatchewan, the native Mixed Grass prairie is dominated by C3 species (Coupland and Rowe, 1969), but there are some C4 grasses as well (Budd et al., 1987; Tober and Chamrad, 1992). Since C3 and C4 forage grasses grow at different rates and in different patterns throughout the growing season, grazing management could alter their growth response and overall productivity in the field (Waller et al., 1994).

Given that C4 grasses grow well in mid-summer, they should be able to support increased grazing livestock production as C3 forage declines in yield and quality. To test this hypothesis, a research study started in 2001 at Swift Current is evaluating the nutritional quality, above-ground biomass production, and grazing performance of two native mixtures: a simple grasses/forb mixture containing cool-season species, and a complex grasses/shrubs/forb mixture containing both cool- and warm-season species. Preliminary results from harvests in August and September indicate that the complex mixture is 20% to 26% greater than the simple mixture in forage protein concentrations and 3% to 8% greater in digestibility concentrations (Iwaasa and Schellenberg, 2003). Average daily gains of beef steers on the complex mixture were consistent during the July to August grazing period, while gains on the simple mixture declined rapidly after July. Other studies (Hall et al., 1982; Ward, 1988; Jackson, 1999) produce similar results, suggesting that the incorporation of C4 grasses into a pasture system can improve cattle weight gains during the summer months compared with grazing only C3 grass pastures. Also, complex native mixtures can provide forage with higher nutritional value that is better able to meet the nutritional needs of ruminants later in the growing season (Cook, 1972). Selective foraging by cattle (preferring some plants and avoiding others) affects the character and composition of rangelands and nutritional quality of the diet (Wallace et al., 1972) and is a possible explanation for seasonal patterns of livestock gains (Hart et al., 1983).

Efficient management of rangelands depends on the identification of plants that are both palatable and nutritious to grazing livestock. Some plant species are relatively unpalatable or not preferred by grazing animals, despite a high nutrient concentration. For example, Hart et al. (1983) found that the forage quality of blue grama (*Bouteloua gracilis* [H B.K.] Lag. ex Steud.) is high in the spring but its quality decreases more rapidly than needle-and-thread (*Stipa comata* Trin. & Rupr.) grass. Yet Vavra et al. (1977) working in Colorado and Samuel and Howard (1982) in Wyoming reported that blue grama ranked the lowest in preference compared to western wheatgrass and needle-and-thread. Even when compared to other warm-season grasses (i.e., big bluestem, little bluestem, side-oats grama, switchgrass and sandreed grass), blue grama was consistently less preferred (Rogler, 1944). However, increased consumption and selection of blue grama was observed as other, more palatable plants become less abundant in the pasture during the progression of the grazing season (Vavra et al., 1977; Samuel and Howard, 1982). Other studies (Caswell et al., 1973; Kautz and Van Dyne, 1978) report that cattle appear to avoid warm-season species and select forbs and cool-season grasses, but Tomanek et al. (1958) found that cattle prefer big and little bluestem grasses over other warm-season grasses and western wheatgrass during the grazing season (May to August). Proper grazing management may encourage the use of such plants (Abouguendia, 1998), and Bai et al. (1998) report that moderate stocking rates tend to favor greater plant species diversity. Presumably, increased diversity would improve the productivity of the pasture and the nutritional status of the grazing animal. Additional research is needed in this area to understand the livestock-plant interface and how to best manipulate cattle behavior to improve rangeland forage production, forage quality, and beef production.

In terms of introduced forage plants, Cruz and Ganskopp (1998) find that crested wheatgrass was selectively grazed by beef steers over native grasses at vegetative and anthesis growth stages. At the same time, some native grass species were preferred over crested wheatgrass at a mature phenological growth stage while others were similar to it. These preferences were observed both in a pasture where crested wheatgrass made up a large proportion of available forage and in a rangeland site where crested wheatgrass was less abundant. The authors concluded that native grasses can provide summer and fall grazing while crested wheatgrass should be used in spring.

The diets of grazing cattle can also be improved by integrating native rangeland into a grazing system. Complementary grazing is currently being promoted in western Canada and may allow pastures to sustain higher stocking rates than continuous, season-long grazing. In this system, cattle are moved through a sequence of pastures with forages that mature in sequence over the season. This gives cattle access to the forage when it has its best quality (Martin and Fredeen, 1999). Crested wheatgrass pastures are grazed in spring while native rangeland pastures are grazed in summer and fall. By integrating native range plants with seeded pasture, producers can delay the grazing of native range and improve the nutrient status of grazing cattle over a longer grazing season (Adams et al., 1996). At the same time, diverse mixtures of seeded, native species can provide improved forage quality and palatability. Combined with managed grazing systems, these mixtures are likely to improve livestock performance and provide an alternative to over-grazing and the potential degradation of our remaining native rangeland resources.

Seed Production

Native grass species exhibit large annual and environmental variation in seed production (Phan and Smith, 2000; Jefferson et al., 2002). This leads to inadequate and inconsistent seed supplies of adapted varieties for many native plant species in the northern Great Plains of western Canada. Phan and Smith (2000) find significant variation for seed yield and seed yield-component traits within collections of blue grama and little bluestem obtained from southern Manitoba. Generally, the most northern collections of both species showed earlier anthesis, produced less biomass, and had lower seed yield than more southern collections when grown at one location. These findings indicate that indigenous plant collections of blue grama and little bluestem show high levels of genetic diversity for seed yield and seed yield components. Genetic diversity is the basis for the development of adapted varieties through directed selection pressure for enhanced seed production capability.

As for native wheatgrasses, northern wheatgrass and western wheatgrass exhibit strong rhizomatous growth, but reproductive tiller density declines over time as vegetative tillers eventually dominate the sward. Slender wheatgrass seed production has been commercialized in western Canada, partly because of high seed yields (Lawrence and Ratzlaff, 1989). High seed production has resulted in low seed costs for this species but can contribute to excessive proportions of slender wheatgrass in native species mixtures for revegetation projects. Slender wheatgrass can dominate native stands for several years after establishment due to its rapid establishment and competitive advantage over other grass species (Hammermeister and Naeth, 1999).

Seed shattering is common in most native plant species as a method of natural seed dispersal. However, it results in a major reduction in both the quantity and quality of seed harvested. The development of methods to improve seed retention may be the most important requirement to enhance the use of native plant species for seeding, since this would lead to improvements in the productivity and quality of seed. However, consideration must be given to the risks of naturalizing native species, such as reduced adaptation, if seed shattering is reduced in varieties of native species.

Most native species have undergone limited, if any, selection for agronomic characteristics, such as increased germination, improved seed processing and handling, and better seedling establishment or reduced seed dormancy (Young and Young, 1986). Baskin and Baskin (1998) note that the seeds of forbs and many native grasses are dormant and that most shrubs have varying forms of seed dormancy and varying lengths of viability once dispersed from the plant. For example, winterfat seed remains viable for only a few months at ambient temperatures. Many native seeds also have structures which increase difficulty of handling. The hulls of white prairie clover (*Petalostemon candidum* [Willd.] Michx.) are difficult to remove and inhibit germination. Stipa spp. have awns as well as barbed tips that often get lodged in seeding and cleaning equipment. Winterfat has hairy bracts which decrease the flow of seed within a seeder. Nuttall's saltbush seed is encapsulated within woody bracts that protect the seed but also decrease imbibition and make it difficult to assess seed size.

Recent research efforts have focused on the collection of a few native plant species in western Canada by government and non-government organizations, such as Agriculture and Agri-Food Canada (AAFC), the Alberta Research Council, Ducks Unlimited Canada, and Prairie Seeds Ltd. Current research at AAFC is aimed at the characterization and further selection and development of a few native plant varieties from collections of ecotypes made within the prairie provinces. These collections are based on populations that captured the maximum genetic diversity from the original ecotypes. While genetically diverse native varieties, called ecovars™, are currently under development (Figure 1) for large scale revegetation projects (Smith and Whalley, 2002), it is unknown whether these varieties have maintained genetic diversity and broad adaptation after selection for seed yield. Additional research is needed to characterize the genetic diversity of the resultant populations, evaluate their range of adaptation across the region, and develop seed production technologies to ensure their commercialization.

Soil Fertility and Carbon

Perennial forages are usually grown on low productivity soils, and ranchers seldom manage fertility for them with the care they do for annual grain crops (Follett and Wilkinson, 1995). The use of chemical fertilizers on rangelands may not be desirable because sustainable production systems are those that rely on minimal inputs of resources, such as chemical fertilizer, to achieve long-term productivity and environmental compatibility (Poincelot, 1987). Consequently, any sustainable production system must depend on efficient nutrient management. Improved nutrient utilization requires greater understanding of the role of biological processes in the release of nutrients from soil organic matter.

Soil organic matter concentration is fundamental to the maintenance of soil fertility because nutrients are critical to plant growth. The rate at which nutrients

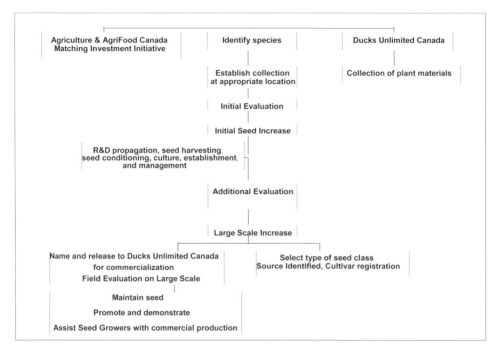

Figure 1. Scheme for development of varieties or ecovars™ of native plant species (adapted from Smith and Whalley, 2002).

are released from soil organic matter is influenced by the chemical composition of the organic matter (including crop residues and root biomass), landscape position, and climate (Gregorich et al., 1995). Plant litter, a layer of dead leaf and stem tissues found at the soil surface, is an important source of carbon input to the soil. In addition, litter is essential in sustaining the prairie ecosystem as a source of energy inputs for soil microbes and as a "sink" for plant nutrients (Wilms et al., 1994). Thus, the characterization and manipulation of soil organic matter (including forage residues and root biomass) concentration and nutrient mineralization rates in forage fields in the prairies are integral to sustainable forage production.

When soils are cultivated for crop production, particularly annual crops, the natural plant-soil system is modified (Gregorich et al., 1995) and soil organic matter concentration decreases. This is due to increased decay of existing soil organic matter as a result of tillage and alterations in soil temperature, water, and aeration as well as changes in the nature of crop residue (straw and root) additions (Swift et al., 1979). The planned conversion from arable agriculture to continuous grass production, either as an introduced species pasture or as a reseeded-native pasture, will result in an increase in soil organic matter levels (Dormaar and Carefoot, 1996), and thereby enhance soil carbon (C) sequestration. The sequestration of C in soil by enhancing soil organic matter (which is mostly C) has been proposed as a "sink" in Kyoto protocol negotiations. During the initial implementation years of the protocol, C sequestration in sinks can be used as offsets against CO_2 emissions (Environment Canada, 2002).

Efforts to meet national targets to decrease C emissions for Kyoto could be aided by planting perennial native plants on land that is marginal for annual crop

Radenbaugh and Sutter: Plate 1.
Organizers for the third "Plain as the Eye See" Public forum (Todd A. Radenbaugh [left] and Glenn C. Sutter [right]), pose in front of an advertisement in front of the Royal Saskatchewan Museum (May 16, 2003).

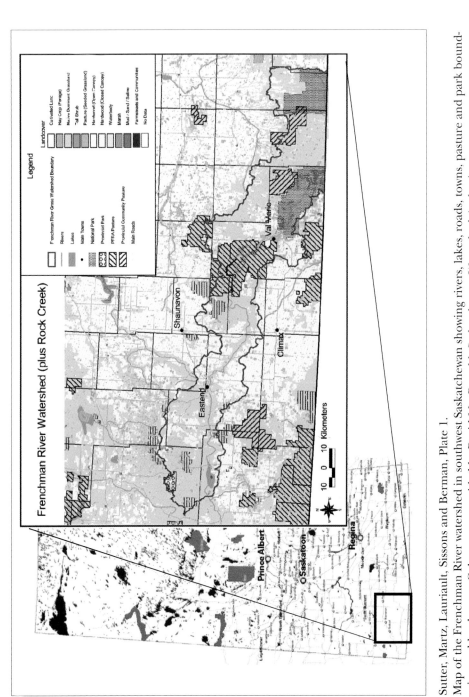

Sutter, Martz, Lauriault, Sissons and Berman, Plate 1.
Map of the Frenchman River watershed in southwest Saskatchewan showing rivers, lakes, roads, towns, pasture and park boundaries, and land cover. Colour map provided by David MacDonald, Saskatchewan Watershed Authority.

Hamilton, Plate 1

Plate1. La Vérendrye's original "prairies" (meadows) must have looked very much like this native oak savanna at Lake Francis Wildlife Management Area, MB on the southeastern side of Lac des Prairies (now Lake Manitoba). Note that the dominant cover of bluestems and prairie dropseed grows less than 0.5 m tall.

Hamilton, Plate 2.

True tall-grass prairie in Kansas has much the same "shaggy" appearance as that illustrated in 1875 (Fig. 1) for the area around Morden, Manitoba.

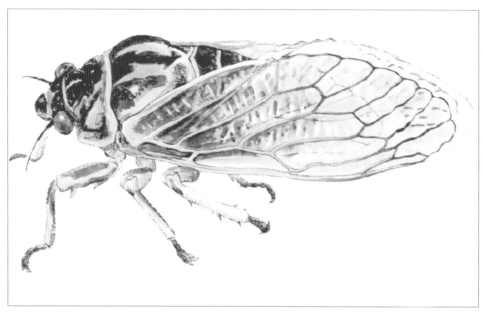

Hamilton, Plate 3.
Okanagana synodica, a boldly coloured cicada (Cicadidae) that is almost invisible in dappled light and shade within a sagebrush bush.

Hamilton, Plate 4.
"Tall-grass" prairie (oak savanna) at Stuartburn, MB after a burn, is dominated by big bluestem growing more than a metre tall; in normal years the growth is less than half as tall.

Hamilton, Plate 5.
Natural prairie openings on bluffs above the Qu'Appelle River east of Round Lake, the easternmost site for several western leafhoppers, and also the westernmost site for some eastern leafhoppers.

Hamilton, Plate 6.
Remnant tall-grass prairie along a roadside just west of Morden, MB has considerable biodiversity due to the slight slope, with bluestems and sunflowers occupying the higher ground, grading off through prairie dropseed and other low plants to cordgrass, sedges and willows.

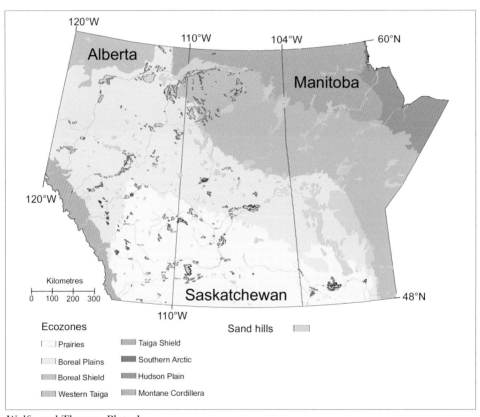

Wolfe and Thorpe, Plate 1.
Distribution of sand hills and ecozones in the Canadian prairie provinces.

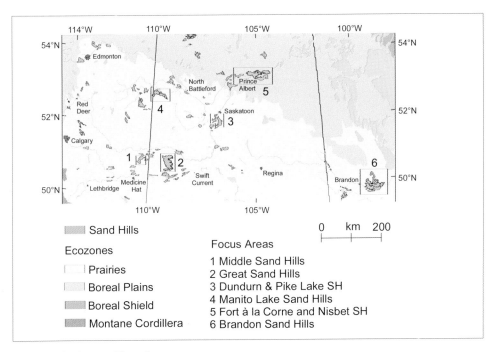

Wolfe and Thorpe, Plate 2.
Focus sand hills and ecozones in the Canadian prairie provinces.

Wolfe and Thorpe, Plate 3.
Active sand dunes along the western edge of the Great Sand Hills (Prairies–mixed grassland ecoregion). Photo by S.A. Wolfe, Sept. 1994.

Wolfe and Thorpe, Plate 4.
Stabilized blowouts and sand hills in the Pike Lake Sand Hills (Prairies–moist mixed grassland ecoregion). Photo by S.A. Wolfe, Oct. 2000.

Wolfe and Thorpe, Plate 5.
Active sand dunes on south-facing slopes in the Manito Lake Sand Hills (Prairies–aspen parkland ecoregion). Photo by S.A. Wolfe, July 2000.

Wolfe and Thorpe, Plate 6.
Sand dunes covered by emergent forest vegetation (in foreground) in the Fort à la Corne Sand Hills following a fire in 1995 and trembling aspen (*Populus tremuloides*) and white spruce (*Picea glauca*) in unburned lowlands (Boreal Plains–boreal transition ecoregion). Photo by S.A. Wolfe, July 2001.

Wolfe and Thorpe, Plate 7.
White spruce forest (*Picea glauca*) with creeping juniper (*Juniperus horizontalis*) on stabilized sand dune in the Brandon Sand Hills (Prairies–aspen parkland ecoregion). Active sand dune is visible in background. Photo by S.A. Wolfe, July 2001.

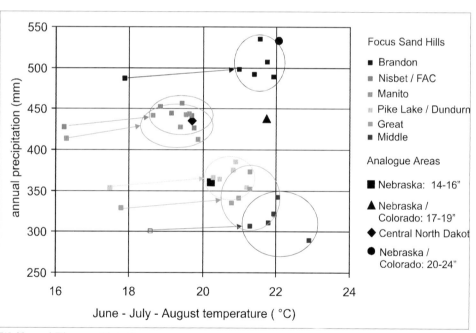

Wolfe and Thorpe, Plate 8.
Mean annual precipitation and summer temperatures under current and future climates of focus sand hills and analogue areas. For each focus area, the single square represents the current climate based on 1961–90 normals. From this, an arrow points to an ellipse that encloses the five scenarios for the 2050s climate. Also shown are the current (1961–90) climates of the US analogue areas.

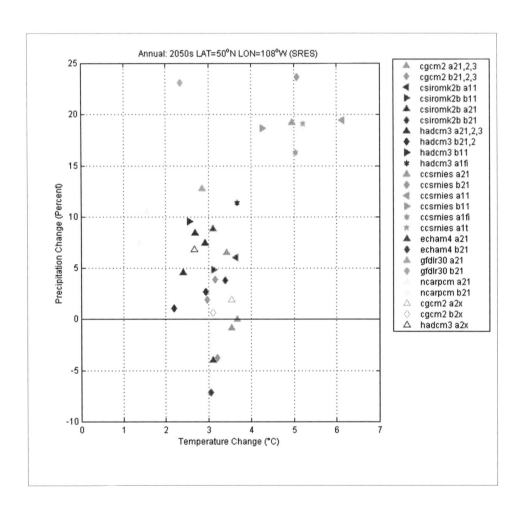

Sauchyn, Kennedy and Stroich, Plate 1.
A suite of climate change scenarios for southern Saskatchewan for the 2050s (2049–60). The forecasted changes in temperature and precipitation are derived from various global climate models and emission scenarios. (see http://www.cics.uvic.ca/scenarios/index.cgi for an explanation of the climate model experiments).

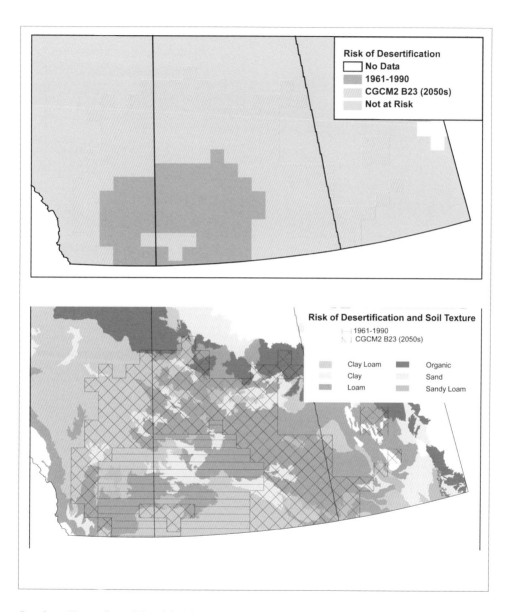

Sauchyn, Kennedy and Stroich, Plate 2.
Maps showing risk desertification (P/PET < 0.65) in the southern Canadian plains for 1961–90 and the 2050s (2049–60). The lower maps include map units classified by soil texture. This map indicates where sandy soils will be subject to dry sub-humid climate and thus risk desertification as the result of climate change.

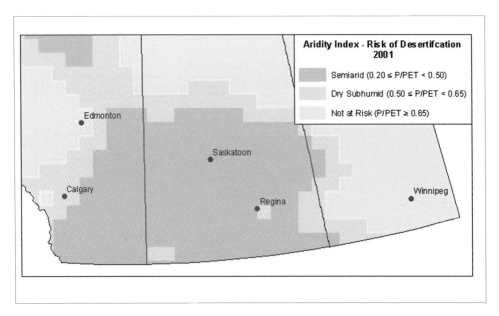

Sauchyn, Kennedy and Stroich, Plate 3.
The aridity index mapped for 2001. A large part of the western Canadian plains had a climate that classifies as semi-arid.

Radenbaugh, Plate 1.
Roads and fence lines subdivide the prairie creating a mosaic of native and agricultural landscapes. Taken by southeast of Wood Mountain, SK.

Radenbaugh, Plate 2.
Woodland and brushland expansion in the northern prairies (A-above); Planted trees in Spring Valley, SK.; (B-below) Picnic area in Dunnet Regional Park, SK; (C-following page, above) Natural growth of brushlands and aspen groves near St Victor, SK; and (D-following page, below) Expansion of brushland and aspen out of small coulee in Grasslands National Park, SK.

production. The Conservation Reserve Program land in the US has greater soil organic C concentration compared to land that was in annual crops (Follett et al., 1999). Soil microbial biomass levels in the 0 to 60 cm depth were 28% higher in CRP soils and 81% higher in native rangeland soil compared to soils from crop-land. Janzen et al. (2000) note that soil organic C in rangeland soil may exceed all above-ground portions of a temperate forest, and this amount can be increased by returning previously-cultivated land back into grassland.

Bremer et al. (1994) find that crested wheatgrass accumulated more total organic soil C and light fraction C than wheat. Smoliak and Dormaar (1985) report that soil from a long-term (23 years) stand of crested wheatgrass had less soil organic C than adjacent native rangeland. Frank et al. (1999) find that native mixed prairie had greater potential as a carbon sink than a monoculture of western wheat-grass. This may have been due to increased root mass and greater exploration of different soil strata by various species. Christian and Wilson (1999) suggest that since total C was less in soils under crested wheatgrass than under native prairie, the planting of crested wheatgrass on millions of hectares of the Great Plains may have left 3.3 to 4.8 x 10^8 tonnes of C in the atmosphere that otherwise would have been stored as soil organic matter by native grass. Thus, if C sequestration is a man-agement goal, then native species should be preferred over introduced forage species for sequestration projects on the prairies.

Grazing stimulates above-ground production, increases tillering and rhizome production, and may stimulate root respiration and root exudation rates (Schuman et al., 2002). All of these factors can result in a change in the amount of C or nitrogen (N) being released or stored in the soil system. Several studies (Schnabel et al., 2001; Schuman et al., 2002) conclude that some grazing manage-ment strategies (for example, grazing intensity) can assist in the rapid incorpora-tion of C into the soil, leading to increased soil organic C (SOC) levels. A study by Schnabel et al. (2001) further concludes that the SOC sequestration potential is considerably higher for lightly and heavily grazed pastures if they are hayed month-ly (1.5 and 1.8 MT C ha-1 year-1 vs 0.3 MT C ha-1 year-1, respectively). Henderson (2000) measured SOC storage in grazed and ungrazed areas at nine native grass-land sites on the southern Canadian Prairies and found that soil C tended to be higher under grazing than in ungrazed exclosures at semi-arid sites (mean annual precipitation of 328 to 390 mm). However, at sub-humid sites (mean annual pre-cipitation of 476 mm) the trend was reversed, suggesting that SOC sequestration depends on factors such as climate, soils, previous management, and potential net primary productivity (Follett, 2003). Canada has about 22 million ha rangeland and improved forage lands and about 5 million ha of cultivated marginal land (CLI 4, 5 & 6) that is economically and environmentally unsustainable across the three western provinces (Smith and Hoppe, 2000). Through the potential reseeding of native forage species, this large land resource could make a great contribution to greenhouse gas reduction by sequestering more C into the soil sink.

Ecological Considerations

Biodiversity

Biodiversity is essential for livestock production on native prairie. Where more species are present, there are more choices of nutritious and palatable forage resources for grazing ruminant livestock (Smoliak and Bezeau, 1967), including later maturing forb or shrub species. Increased biodiversity is also associated with increased ecosystem productivity (Tilman et al., 1996) and primary production that is more resistant to drought stress (Tilman and Downing, 1994). Current research suggests that a more diverse and complex mixture of cool and warm season grasses combined with shrub and forb species will produce more steer gains in late summer compared to a simple mix of cool-season grasses (Iwaasa and Schellenberg, 2003). Mixed prairie is the forage of choice for steers in August and September, as opposed to a monoculture of crested wheatgrass (Schellenberg et al., 1999). In Nevada, the proportion of shrub species, including winterfat, in revegetation seed mixtures was higher when biodiversity and habitat restoration were primary objectives, compared to mixtures seeded for watershed protection and soil erosion control (Richards et al., 1998). These authors conclude that the cost of diverse native seed mixtures frequently restricts their use, even by government agencies or non-government organizations whose policies promote the use of native species.

Increased plant biodiversity can decrease resource limitations such as N through biological N fixation (Schellenberg and Banerjee, 2002) thus increasing forage productivity. Wilson (2000) finds that increased biodiversity can also increase the heterogeneity of abiotic resources. Heterogeneity of nutritional resources often varies over time and space, and small differences may have impacts on plant physiology and competition (Bakker, 2000). Increased biodiversity can result in several levels of competition for soil nutrient resources (Wilson, 2000). Davis (2003) suggests that the availability of soil nutrient resources is often the deciding factor in the success of invasive species. It follows that diverse mixtures of reseeded native species may be less susceptible to invasion by exotic weeds, but this hypothesis needs to be tested.

Deciding between native or introduced species in rangeland seeding is a multiple component process that requires the consideration of philosophy, objectives, site potential and limitations, availability and cost of plant materials, weed invasion, desired community seral stage, and economic limitations (Jones and Johnson, 1998). Critics suggest that seeding varieties or ecovars™ of native species introduces genetic change into natural landscapes but several naturally-occurring interspecific grass hybrids are evidence that non-anthropogenic genetic change and intra-specific diversity can occur in native grasslands. While genetically diverse populations or localized varieties of native species would be ideal for revegetation, such ideals are often unavailable or too expensive for large-scale projects (Jones and Johnson, 1998). The best decisions require knowledge about the genetic variation within an individual species, how this variation relates to ecological adaptations, and interactions that link the species to other ecosystem components.

Climate Change

In the near future, climate change will impact crop species and their productivity in the current agricultural regions of the US (Antle, 1997). World Resources Institute (1990) predicts that on the northern Great Plains of North America, the mean May to August temperatures will rise by 3.5°C over the next century. The frost-free growing season is also expected to lengthen by four to nine weeks with an increased frequency of drought in July and August. Environment Canada (2001) projects a similar 3°C annual temperature increase for the prairie region of western Canada. Cutforth et. al (1999) examine long-term (50 years) weather records for 15,000 km² of semiarid prairie southwest of Swift Current, Saskatchewan, and find that winter and spring maximum and minimum temperatures have increased and precipitation has decreased. Climate change has been linked to plant species migration (Root et al., 2003) and to shifts in species ranges (Higgins et al., 2003). As the climate warms up, plant communities are likely to shift in favour of species that tolerate water and high temperature stress.

Pastures and rangelands are thought to be very sensitive to climate change (Gregory et al., 1999). Climate warming has already led to a decline in species richness (Sala et al., 1999), and future changes in plant species composition may have large impacts on livestock production capabilities due to changes in nutritional quality and time of availability (Gregory et al., 1999). Climate warming over several centuries in southwestern Saskatchewan has favoured C3 species over C4 species at Grasslands National Park, located in the semi-arid prairie region near Val Marie, Saskatchewan (Peat, 1997). This result is not surprising given the biochemical advantage of C3 photosynthesis compared to C4 photosynthesis at ambient CO_2 concentrations. Analysis of long term vegetation data in Colorado (Alward et al., 1999) find a similar shift in production in favour of C3 species over several decades. As climate warming progresses, native species may provide useful genetic resources to respond to future climates of the northern Great Plains. Further research is required to elucidate the potential economic effects of climate change on livestock production and society's appreciation of grassland landscapes.

Bio-invasion

Reseeding native species for forage and grazing may be preferable to seeding introduced forage species because some introduced forage species are invasive, posing a threat to neighbouring native plant communities. Bio-invasion involves the replacement of native species by exotic species, which are usually introduced to the environment through human activity. Invasion by exotic species is a powerful driver of global change (Sala et al., 1999), and both climate change and exotic plant invasion contribute to habitat fragmentation and reduced biodiversity. As many as 80% of the world's endangered species are threatened due to the competitive pressures of introduced or exotic species (Armstrong, 1995). During European settlement of Australia, 466 plant species were introduced for hay and/or pasture production, but only 21 species showed merit and 17 of these were deemed to be invasive of adjacent native rangeland (Bright, 1998). Davis (2003) states that no introduced plant species has caused the extinction of another plant species, but some may have contributed to extinction via competition-mediated reduction in fitness. The primary cause of extinction is habitat loss, which can be associated with climate change and bio-invasion working together with human development pressure.

Davis (2003) concludes that 4000 plant species have been introduced into North America. Most were introduced for food, fibre, and ornamentals (Pimentel et al., 2000), with some species accounting for 98% of the food production in the US. Among the introduced forage species, D'Antonio and Vitousek (1992) conclude that invasion by alien grass species is most severe in the arid and semi-arid west of the U.S.A. Invasions by annual grasses were largely unplanned, while many perennial invasive grasses were purposely introduced for livestock forage or to prevent soil erosion (D'Antonio and Vitousek, 1992). The use of introduced forage grass species for pastures and erosion control has led to the widespread distribution and dominance of several species throughout North America (Haber, 1996). Smoliak and Dormaar (1985) estimate that 1 million ha have been seeded to crested wheatgrass in the Canadian prairie provinces alone, and this area has undoubtedly increased since their work was published. White et al. (1993) consider smooth bromegrass (*Bromus inermis* Leyss.), yellow sweetcclover (*Melilotus officinalis* (L.) Pall.), and white sweetclover (*Melilotus alba* Medic.) to be moderately invasive, while alfalfa, crested wheatgrass, and Kentucky blue grass (*Poa pretensis* L.) are minor invasive species of uplands. All of these species were introduced for pasture or hay production in Canada.

The proportion of exotic flora was estimated to be 16% on the prairies of western Canada (Haber, 1996) and 20% for North America (Davis, 2003). In Saskatchewan, smooth bromegrass and crested wheatgrass are considered to be invasive in native fescue grassland (Romo and Grilz, 1990). Looman and Heinrichs (1973) find that crested wheatgrass pasture was not invaded by native plant species until stands were more than 15 years old, and even when native species were found in old crested wheatgrass stands, their contribution to above-ground biomass did not exceed 10% of the total annual forage production. Thus, seeded monocultures of crested wheatgrass appear to resist successional species change, even under heavy grazing. Similar results were found in crested wheatgrass pastures of southern Alberta (Smoliak et al., 1967). To date, the risk of bio-invasion has not been systematically evaluated for most introduced forage species of the semi-arid prairie.

Caution must be used in recommending restrictions on the use of these introduced forage-grass species. A large proportion of the range livestock industry of the northern Great Plains depends on these species and economic losses could occur if a wholesale move to native species only was forced by legislation or government policy. There is not sufficient seed for such action, and more research is necessary to assess the capabilities of native species varieties. To date, only a relative few native species have been examined and some lack even the basic agronomic information. Moreover, large areas were seeded with introduced forage species to reduce erosion (Gray, 1996) and removal of the vegetative cover they provide without an adequate plan for replacement with native species could result in environmental disaster. The issue of exotic bio-invasion in Canada lacks practical action on the Government's part (Commissioner of the Environment and Sustainable Development, 2002), though efforts are being made to address the issue within Agriculture and Agri-Food Canada's research portfolio. Management of invasive introduced species, their impact on native plant communities in agricultural and environmental terms, and the invasive potential of native species require further research.

Management Implications

The use of native versus introduced species for forage production and grazing has been a point of controversy for decades on the northern Great Plains. A new paradigm for selecting plant species was proposed by Brown and Amacher (1999). They suggest that introduced or exotic species are those which dominate early in ecological succession (early seral) and are characterized by high biomass yield, fast growth rates, high seed production, aggressive growth habit, and responsiveness to soil fertility. Native species can be characterized as those which dominate late successional (late seral) stage communities and exhibit slower growth rate, variable seed production, compatibility with other species, and adaptation to low soil fertility. Based on this model, all species would be considered depending on the goals and objectives of the restoration project. Our review has shown that both native and introduced species have distinct advantages depending on their intended use.

Native plant species will contribute to sustainable agricultural systems of the new century in the semi-arid prairie. We conclude that more research is needed on several topics. The genetic diversity of new native plant varieties needs to be established and the value of intra-variety genetic diversity assessed relative to intensively selected varieties. In addition, the geographic range of adaptation of native varieties and their role in summer and fall grazing by beef cattle needs to be determined, and seed production technologies that will permit the efficient commercialization of these species need to be studied. For some species, this may be as simple as documenting techniques developed by individual producers; for others, research into seed dormancy will be required. The establishment of shrub species to improve rangelands for fall and winter grazing also needs further study. Second, changes in soil organic C after seeding native species on previously cropped land need to be quantified, including the impact of species type and grazing management on the rate and extent of C sequestration.

Acknowledgements

The authors thank the Agriculture Development Fund of Saskatchewan, Ducks Unlimited Canada, Horned Cattle Trust Fund of Saskatchewan, Canada-Saskatchewan Green Plan, Saskatchewan Stock Growers Association, and Agriculture and Agri-Food Canada's Matching Investment Initiative for funding research related to this topic. Technical assistance from Russell Muri, Ed Birkedal, Jacqueline Bolton, Elizabeth Chan, Darwyn Wilms, Trent Olson, Perry Coward, Karen Tsougrianis, Cliff Ratzlaff, and Vivian Boyer is greatly appreciated. Thought-provoking discussions on this topic with Thomas A. Jones, Neal W. Holt, J. Waddington, Brent Wark, and Thomas O. Dill contributed greatly to the development of this article.

References

Abouguendia, Z. 1998. *Nutrient Content and Digestibility of Saskatchewan Range Plants: A Summary Report.* GAPT (July). 3

Adams, D.C., R.T. Clark, T.J.Klopfenstein and J.D. Volesky. 1996. "Matching the Cow with Forage Resources," *Rangelands* 18: 57–62. 5

Agriculture and Agri-Food Canada. 2003. Greencover Canada. Online: www.agr.gc.ca/env/greencover-verdir [accessed March 16, 2004].

Alderson J. and W.C. Sharp. 1994. *Grass Varieties in the United States.* Handbook No. 170 Washington, DC: USDA-ARS.

Alward, R.D., J.K. Detling and D.G. Milchunas. 1999. "Grassland Vegetation Changes and Nocturnal Global Warming," *Science* 283: 229–231.

Antle, J.M. 1997. *Climate Change and Economic Constraints Facing Great Plains Agriculture.* Briefing Document for Great Plains Regional Climate Change Workshop. Loveland, CO, May 27–29, 1999. In *Proceedings Climate Change Impacts on Rangeland Systems Symposium and Workshop*, SRM Meeting, Feb. 25–26, 1999. Omaha, NE.

Armstrong, S. 1995. "Rare Plants Protect Cape's Water Supplies," *New Scientist* (February 11): 8.

Arthun, D., J.L. Holechek, J.D. Wallace, M.L. Gaylean and M. Cardenas. 1992. "Forb and Shrub Effects on Ruminal Fermentation in Cattle," *J. Range Manage.* 45: 519–522.

Asay, K.H., W.H. Horton, K.B. Jensen and A.J.Palazzo. 2001. "Merits of Native and Introduced Triticeae Grasses on Semiarid Rangelands," *Can. J. Plant Sci.* 81: 45–52.

Bai, Y., Z. Abouguendia and R.E. Redmann. 1998. "Effect of Grazing on Plant Species Diversity of Grasslands in Saskatchewan." Proc. Of 5th Prairie Conservation and Endangered Species Conf. Feb. 19–22, Saskatoon, SK.

Bakker, J.P. 2000. "Environmental Heterogeneity: Effects on Plants in Restoration Ecology." Pp. 379–400 in M.J. Hutchings, E.J. John and A.J.A. Stewart (eds.), *The Ecological cConsequences of Environmental Heterogeneity.* Oxford: Blackwell Science.

Banerjee, M.R. and M.P. Schellenberg. 2000. "Legume and Native Shrub Mixtures for Potential Optimum Forage Production." Proc. Soils and Crops Workshop. Feb. 24–25, 2000. Saskatoon, SK.

Baskin, C.C. and J.M. Baskin. 1998. *Seeds: Ecology, Biogeography, and Evolution of Dormancy and Germination.* San Diego: Academic Press.

Bremer, E., H.H. Janzen and A.M. Johnston. 1994. "Sensitivity of Total, Light Fraction and Mineralizable Organic Matter to Management Practices in a Lethbridge Soil." *Can J. Soil Sci.* 74: 131–138.

Bright, C. 1998. *Life Out of Bounds: Bioinvasion in a Borderless World.* Worldwatch Environmental Alert Series. New York: W.W. Norton & Company.

Brown, R.W. and M.C. Amacher. 1999. "Selecting Plant Species for Ecological Restoration: A Perspective for Land Managers." Pp. 1-16 in: L.K. Holzworth and R.W. Brown (eds.), *Revegetation with Native Species.* Proceedings, 1997 Society for Ecological Restoration annual meeting; PMRS-P-8, United States Department of Agriculture, Forest Service,Rocky Mountain Research Station, Ogden, UT.

Budd, A.C., J. Looman and K.F. Best. 1987. *Budd's Flora of the Canadian Prairie Provinces.* Ottawa: Research Branch, Agriculture Canada Publication 1662.

Caswell, H., F. Reed, S.N. Stephenson and P.A.Werner. 1973. "Photosynthetic Pathways and Selective Herbivory: A Hypothesis." *American Naturalist* 107: 465–481.

Christian, J.M. and S.D. Wilson. 1999. "Long-term Ecosystem Impacts of an Introduced Grass in the Northern Great Plains." *Ecology* 80: 2397–2407.

Commissioner of the Environment and Sustainable Development. 2002. Annual Report: Ch. 4, "Invasive Species." Http://www.oag-bvg.gc.ca/domino/reports.nsf/html/c2002004ce.html (accessed May 30, 2003).

Cook, C.W. 1972. "Comparative Nutritive Values of Forbs, Grasses and Shrubs." Pp 303–310 in C.M. McKell, J.P. Blaisdell and J.R. Goodin (eds.), *Wildland Shrubs: Their Biology and Utilization.* USDA Forest Service GTR INT-1.

Cordova, F.R. and J.D. Wallace. 1975. "Nutritive Value of Some Browse and Forb Species," *West. Sec. Amer. Soc. Anim. Sci.* 26: 160–162.

Coupland, R.T. 1974. Matador Project. Technical Rep. No. 62. Can. Comm. for the Int. Biol. Progr. University of Saskatchewan, Saskatoon, Sask.

Coupland, R.T. and J.T. Rowe. 1969. "Natural Vegetation of Saskatchewan." Pp. 73–78 in: J.H. Richards and K.I. Fung (eds.), *Atlas of Saskatchewan.* Saskatoon: Modern Press. 16

Cruz, R. and D. Ganskopp. 1998. "Seasonal Preferences of Steers for Prominent Northern Great Basin Grasses, *J. Range Manage.* 51: 557–565.

Cutforth, H.W., B.G. McConkey, R.J. Woodvine, D.G. Smith, P.G. Jefferson, and O.O. Akenremi. 1999. "Climate Change in the Semiarid Prairie of Southwestern Saskatchewan: Late Winter-Early Spring," *Can.J. Plant Sci.* 79: 343–350.

D'Antonio, C.M. and P.M. Vitousek. 1992. "Biological Invasions by Exotic Grasses, the Grass/Fire Cycle and Global Change," *Annual Review of Ecology and Systematics* 23: 63–87.

Davis, M.A. 2003. "Biotic Globalization: Does Competition from Introduced Apecies Threaten Biodiverity?" *Bioscience* 53: 481–491.

Dormaar, J.F. and J.M. Carefoot. 1996. "Implications of Crop Residue Management and Conservation Tillage on Soil Organic Matter," *Can. J. Plant Sci.* 76: 627–634.

Dubbs, A.L., R.T. Harada and J.R. Stroh. 1974. "Evaluation of Thickspike Wheatgrass for Dryland Pasture and Range," *Montana State University Bulletin* 677.

Environment Canada. 2001. "The Science of Climate Change." Online: www.ec.gc.ca/climate/overview_science-e.html [accessed March 31, 2004].

——. 2002. "Climate Change Plan for Canada. Achieving Our Commitments Together. F. Agriculture, Forestry and Landfills. Online: www.climatechange.gc.ca/plan_for_canada/plan/chap_3_6.html [accessed March 16, 2004].

Follett, R.F. 2003. "Pasture Management Systems: Impact on Soil Carbon and Greenhouse Gases." *Proceedings of 2003 CSAS Annual Meeting, June 10–13, 2003.* University of Saskatchewan.

Follett, R.F., E.G. Pruessner, J.M. Kimble, S.E. Samson-Liebig, and S.W. Leavitt. 1999. "Soil-C 17 Storage Within Soil Profiles of the Historical Grasslands of the USA." Briefing Document for Great Plains

Regional Climate Change Workshop, Loveland, CO, May 27–29, 1999. In *Proceedings Climate Change Impacts on Rangeland Systems Symposium and Workshop*, SRM Meeting, Feb. 25–26, 1999. Omaha, NE.

Follett, R.F. and S.R.Wilkinson. 1995. "Nutrient Management of Forages." Pp. 55–82 in R.F. Barnes, D.A. Miller and C.J.Nelson (eds.), *Forages Vol. II: The Science of Grassland Agriculture*. Ames: Iowa State University Press.

Frank, A.B. and A. Bauer. 1991. "Rooting Activity and Water Use during Vegetative Development of Crested and Western Wheatgrass," *Agron. J.* 83: 906–910.

Frank, A.B., W.A. Dugas, J.F. Karn and H.S. Mayeux. 1999. "Carbon Dioxide Fluxes over Grazed Grasslands." Pp. 214–215 in D. Eldridge and D. Freundenberger (eds.), *Proceedings of the VI 6 International Rangeland Congress; People and Rangelands: Building the Future*. Townsville, Australia, July 19–23, 1999.

Gebhart, D.L., H.B. Johnson, H.S. Mayeux and H.W. Polley. 1994. "The CRP Increases Soil Organic Carbon," *J. Soil and Water Cons.* 49: 488–492.

Gray, J.H. 1996. *Men Against the Desert*. Saskatoon: Western Producer Prairie Books.

Gregorich, E.G., D.A. Angers, C.A. Campbell, M.R. Carter, C.F. Drury, B.H. Ellert, P.H. Groenevelt, D.A. Holmstrom, C.M. Monreal, H.W. Rees, R.P. Voroney and T.J. Vyn. 1995. "Changes in Soil Organic Matter." In D.F. Acton and L.J. Gregorich (eds.), *The Health of Our soils: toward Sustainable Agriculture in Canada*. Ottawa: Agriculture and Agrifood Canada Publication #1906/E.

Gregory, P.J., J.S.I. Ingram, B. Campbell, J. Goudriaan, L.A. Hunt, J.J. Landsberg, S. Linder, S.S. Smith, R.W. Sutherst and C. Valentin. 1999. "Managed Production Systems." Pp. 229–270 in: B. Walker, W. Steffen, J. Canadell and J. Ingram (eds.), *The Terrestrial Biosphere and Global Change: Implications for Natural and Managed Ecosystems*. Cambridge: Cambridge University Press.

Haber, E. 1996. Invasive Exotic Plants of Canada Fact Sheet No. 1. Online: www.infoweb.magi.com/~ehaber/fact1.html. (Accessed February 17, 2000).

Hall, K.E., J.R. George and R.R. Riedl. 1982. "Herbage Dry Matter Yields of Switchgrass, Big Bluestem and Indiangrass with N Fertilization," *Agronomy* J. 74: 47–51.

Hammermeister, A. and A. Naeth. 1999. "An Ecological Approach to Seed Mix Design and Native Prairie Revegetation." Pp. 55–60 in J. Thorpe, T.A. Steeves, and M. Gollop (eds.), *Proc. of Fifth Prairie Conservation and Endangered Species Conference*, Provincial Museum of Alberta, Natural History, Occasional Paper No. 24. Edmonton.

Hanson, C.L., G.A. Schumaker and C.J. Erickson. 1976. "Influence of Fertilization and Supplemental Runoff Water on Production and Nitrogen Content of Western Wheatgrass and Smoothbrome," J. *Range Manage.* 29: 406–409.

Hart, R.H., O.M. Abdalla, D.H. Clark, M.B. Marshall, M.H. Hamid, J.A. Hager, and J.W. Waggoner Jr. 1983. "Quality of Forage and Cattle Diets on the Wyoming High Plains," *J. Range Manage.* 36: 46–51.

Hart, R.H., M.J. Samuel, P.S. Test and M.A. Smith. 1988. "Cattle, Vegetation, and Economic Responses to Grazing Systems and Grazing Pressure," *J. Range Manage.* 41: 282–286.

Henderson, D.C. 2000. "Carbon Storage in Grazed Prairie Grasslands of Alberta." M.Sc. thesis, University of Alberta.

Higgins, S.I., J.S. Clark, R. Nathan, T. Hovestadt, J.M. Schurr, J.M.V. Fragoso, M.R. Aguiar, E. Ribbens and S. Lavorel. 2003. "Forecasting Plant Migration Rates: Managing Uncertainty for Risk Assessment," *J. Ecology* 91: 341–347.

Hofmann, L., R.E. Ries, J.F. Karn and A.B. Frank. 1993. "Comparison of Seeded and Native Pastures Grazed from Mid-May through September," *J. Range Manage.* 46: 251–254.

Holechek, J.L. and C.H. Herbel. 1986. "Supplementing Range Livestock," *Rangelands* 8: 29–33.

Holechek, J.L., R.D. Pieper and C.H. Herbel. 1989. *Range Management Principles and Practices*. Englewood Cliffs, NJ: Regents/Prentice Hall.

Iwaasa, A.D. and M.P. Schellenberg. 2003. "Progress Report on Re-establishment of a Mixed Native Grassland in Southwest Saskatchewan ADF Project #20010042.

Jackson, L.L. 1999. "Establishing Tallgrass Prairie on Grazed Permanent Pasture in the Upper Midwest," *Restoration Ecology* 7: 127–138.

Janzen, H.H., B.H. Ellert and J.F. Dormaar. 2000. "Rangelands and the Global Carbon Cycle." In *Proceedings of The Range: Progress and Potential*, Jan. 23–25, 2000, Lethbridge, AB.

Jefferson, P.G., N.W. Holt and E. Birkedal. 1997. "Use of Seeded Native Grasses for Beef Production." Saskatchewan Agriculture Development Fund Project no. 95000086.

Jefferson, P.G., W.P. McCaughey, K. May, J. Woosaree, and L. McFarlane. 2004. "Forage Quality of Seeded Native Grasses in the Fall Season on the Canadian Prairie Provinces." *Can. J. Plant Sci.* 84: 503–509.

Jefferson, P.G., W.P. McCaughey, K. May, J. Woosaree, L. McFarlane, and S.M.B. Wright. 2002. "Performance of American Native Grass Cultivars in the Canadian Prairie Provinces," *Native Plants Journal* 3: 24–33.

Jefferson, P.G., L. Wetter and B. Wark. 1999. "Quality of Deferred Forage from Waterfowl Nesting Sites on the Canadian Prairies," *Can. J. Anim. Sci.* 79: 485–490.

Johnston, A. 1987. *Plants and the Blackfoot.* Lethbridge, AB: Graphcom Printers Ltd.

Johnston, A., A.D. Smith, L.E. Lutwick, S. and Smoliak. 1968. "Fertilizer Response of Native and Seeded Ranges," *Can. J. Plant Sci.* 48: 467–472.

Jones, T.A. and D.A. Johnson. 1998. "Integrating Genetic Concepts into Planning Rangeland Seedings," *J. Range Manage.* 51: 594–606.

Kamstra, L.D. 1973. "Seasonal Changes in Quality of Some Important Range Grasses," *J. Range Manage.* 26: 289–291.

Karn, J.F. and R.J. Lorenz. 1983. "Supplementation of Yearling Steers Grazing Fertilized and Unfertilized Northern Plains Rangeland," *J. Range Manage.* 36: 41–45.

Kautz, J.E. and G.M. Van Dyne. 1978. "Comparative Analyses of Diets of Bison, Cattle, Sheep and Pronghorn Antelope on Shortgrass Prairie in Northeastern Colorado." Pp. 438–443 in D.N. Hyder (ed.), *Proc. First International Rangelands Congress.* Denver, CO.

Kilcher, M.R. and J. Looman. 1983. "Comparative Performance of Some Native and Introduced Grasses in Southern Saskatchewan, Canada," *J. Range Manage.* 36: 654–657.

Knowles, R.P. 1987. "Productivity of Grass Species in the Dark Brown Soil Zone of Saskatchewan," *Can. J. Plant Sci.* 67: 719–725.

Lawrence, T. 1978. "An Evaluation of Thirty Grass Populations as Forage Crops for Southwestern Saskatchewan," *Can. J. Plant Sci.* 58: 107–115.

Lawrence, T. and C.D. Ratzlaff. 1989. "Performance of Some Native and Introduced Grasses in a Semiarid Region of Western Canada," *Can. J. Plant Sci.* 69: 251–254.

Lawrence, T. and J.E. Troelsen. 1964. "An Evaluation of 15 Grass Species as Forage Crops for Southwestern Saskatchewan," *Can. J. Plant Sci.* 44: 301–310.

Lewis, W.H. and M.P.F. Elvin-Lewis. 2003. *Medical Botany. Plants Affecting Human Health.* Hoboken, NJ: John Wiley & Sons, Inc.

Looman, J. and D.H. Heinrichs 1973. "Stability of Crested Wheatgrass Pastures under Long-term Pasture Use," *Can. J. Plant Sci.* 53: 501–506.

Martin, R. and A. Fredeen. 1999. "Effect of Management of Grasslands on Greenhouse Gas Balance." Agriculture and Agri-Food Canada Report Sept. 1999.

McGowan, D.C. 1975. *Grassland Settlers: The Swift Current Region During the Era of the Ranching Frontier.* Regina: Canadian Plains Research CenteR.

Milchunas, D.G., M.I. Dyer, O.C. Wallmo and D.E. Johnson. 1978. "*In vivo-in vitro* Relationships of Colorado Mule Deer Forages." Colorado Div. Wildlife Spec. Rep. 40, Fort Collins, CO.

Native Plant Working Group. 2001. *Native Plant Revegetation Guidelines for Alberta.* Edomonton: Alberta Agriculture, Food, and Rural Development and Alberta Environment.

Paoletti, M.G., D. Pimentel, B.R. Stinner and D. Stinner. 1992. "Agroecosystem biodiversity: Matching Production and Conservation Biology." Pp. 3–24 in M.G. Paoletti and D. Pimentel (eds.), *Biotic Diversity in Agroecosystems.* Amsterdam: Elsevier.

Peat, H. 1997. "Dynamics of C3 and C4 Productivity in Northern Mixed Grass Prairie." M.Sc. thesis, University of Toronto.

PCAP Partnership. 2003. S*askatchewan Prairie Conservation Action Plan 2003–2008.* Regina: Canadian Plains Research Center.

Phan, A.T. and S.R. Smith Jr. 2000. "Seed Yield Variation in Blue Grama and Little Bluestem Plant Collections in Southern Manitoba, Canada," *Crop Sci.* 40: 555–561.

Pigden, W.J. 1953. "The Relation of Lignin, Cellulose, Protein, Starch and Ether Extract to the Curing of Range Grasses," *Canadian J. Agricultural Science* 33: 364–378

Pimentel, D., L. Lach, R. Zuniga and D. Morrison. 2000. "Environmental and Economic Costs Associated with Non-indigenous Species in the United States." Pp. 285–306 in D. Pimentel (ed.), *Biological Invasions:Economic and Environmental Costs of Alien Plant, Animal and Microbe Species.* Boca Raton, FL: CRC Press.

Poincelot, R.P. 1987. *Toward a More Sustainable Agriculture.* Westport, CT: AVI Publishing Co.

Richards, R.T., J.C. Chambers and C. Ross. 1998. "Use of Native Plants on Federal Lands: Policy and Practice," *J. Range Manage.* 51: 625–632.

Rogler, G.A. 1944. "Relative Palatabilities of Grasses under Cultivation on the Northern Great Plains," *J. Amer. Soc. Agronomy* 36: 487–497.

Romo, J.T., D.T. Booth and C. Zabek. 1997. "Seedbed Requirements and Cold Tolerance of Winterfat Seedlings: An Adapted Forage for the Canadian Prairies." Final ADF Report, Regina, SK.

Romo, J.T. and P.L. Grilz. 1990. "Invasion of the Canadian Prairies by an Exotic Perennial," *Blue Jay* 48: 131–135.

Root, T.L., J.F. Price, K.R. Hall, S.H. Schneider, C. Rosenzweig and J.A. Ponds. 2003. "Fingerprints of Global Warming on Wild Animals and Plants," *Nature* 421: 57–60.

Sala, O.E., F.S. Chapin III, R.H. Gardner, W.K. Laurenroth, H.A. Mooney and P.S. Ramakrishnan. 1999. "Managed Production Systems." Pp. 304–328 in B. Walker, W. Steffen, J. Canadell and J. Ingram (eds.), *The Terrestrial Biosphere and Global Change: Implications for Natural and Managed Ecosystems.* Cambridge : Cambridge University Press.

Samson, F.B., F.L. Knopf and W.R. Ostlie. 1998. "Grasslands." Pp. 437–472 in M.J. Mac, P.A. Opler, C.E. Puckett, and P.D. Doran (eds.),*Status and Trends of the Nation's Biological Resources*, Vol. 2. Reston, VA: U.S. Department of the Interior, U.S. Geological Survey.

Samuel, M.J. and G.S. Howard. 1982. "Botanical Composition of Summer Cattle Diets on the Wyoming High Plains," *J. Range Manage.* 35: 305–308.

Schellenberg, M.P. and M.R. Banerjee. 2002. "Looking for a Legume-Shrub Mixture for Optimum Forage Production: A Greenhouse Study," *Can. J. Plant Sci.* 82: 357–363.

Schellenberg, M.P., N.W. Holt and J. Waddington. 1999. "Effects of Grazing Dates on Forage and Beef Production of Mixed Prairie Rangeland," *Can. J. of Anim. Sci.* 79: 335–341.

Schellenberg, M.P. and P.G. Jefferson. 1998. "Forage for Late Season Grazing." *SPARC, Research Newsletter* 5 (March 27).

Schnabel, R.R., A.J. Franzluebbers, W.L. Stout, M.A. Sanderson, M.A. and J.A. Studemann. 2001. "The Effects of Pasture Management Practices." Pp. 291–322 in R.F. Follett, J.M. Kimble and R. Lal (eds.), *The Potential of US Grazing Lands to Sequester Carbon and Mitigate the Greenhouse Effect.* Boca Raton, FL: Lewis Publishers.

Schuman, G.E., H.H. Janzen and J.E. Herrick. 2002. "Soil Carbon Dynamics and Potential Carbon Sequestration by Rangelands," *Environmental Pollution* 116: 391–396.

Selby, C.J. and M.J. Santry. 1996. *A National Ecological Framework for Canada: Data Model, Database, and Programs.* Ottawa: Agriculture and Agri-Food Canada and Environment Canada.

Smith, D.G. and T.A. Hoppe. 2000. *Prairie Agricultural Landscapes: A Land Resource Review.* Regina: Agriculture and Agri-Food Canada, Prairie Farm Rehabilitation Administration.

Smith, S.R. Jr. and R.D.B. Whalley. 2002. "A Model for Expanded Use of Native Grasses," *Native Plants Journal* 3: 38–49.

Smoliak, S. and L.M. Bezeau. 1967. "Chemical Composition and *in vitro* Digestibility of Range Forage Plants of the *Stipa-Bouteloua* Prairie," *Can. J. Plant Sci.* 47: 161–167.

Smoliak, S. and J.K. Dormaar. 1985. "Productivity of Russian Wildrye and Crested Wheatgrass and Their Effect on Prairie Soils," *J. Range Manage.* 38: 403–405.

Smoliak, S., A. Johnston, and L.E. Lutwick. 1967. "Productivity and Durability of Crested Wheatgrass in Southeastern Alberta," *Can. J. Plant Sci.* 47: 539–548.

Smoliak, S. and S.B. Slen. 1974. "Beef Production on Native Range, Crested Wheatgrass and Russian Wildrye Pastures," *J. Range Manage.* 27: 433–436.

Solutions 2000+. 1997. "Market Assessment of Native Plant Materials in Saskatchewan." Native Plant Society of Saskatchewan and Northwest Saskatchewan Grasslands Association.

Swift, M.J., O.W. Heal and J.M. Anderson. 1979. *Decomposition in Terrestrial Ecosystems.* Berkeley: University of California Press.

Tilman, D. and J.A. Downing. 1994. "Biodiversity and Stability in Grasslands," *Nature* 367: 363–365.

Tilman, D., D. Wedin and J. Knops. 1996. "Productivity and Sustainability Influenced by Biodiversity in Grassland Ecosystems," *Nature* 379: 718–720.

Tober, D.A. and A.D. Chamrad. 1992. "Warm-season Grasses in the Northern Great Plains," *Rangelands* 14: 227–230.

Tomanek, G.W., E.P. Martin and F.W. Albertson. 1958. "Grazing Preference Comparisons of Six Native Grasses in the Mixed Prairie," *J. Range Manage.* 11: 191–193.

Vavra, M., R.W. Rice, R.M. Hansen and P.L. Sims. 1977. "Food Habits of Cattle on Shortgrass Range in Northeastern Colorado," *J. Range Manage.* 30: 261–263.

Wallace, J.D., J.C. Free and A.H. Denham. 1972. "Seasonal Changes in Herbage and Cattle Diets on Sand Hills Grassland," *J. Range Manage.* 25: 100–104.

Waller, S.S., L.E.Moser and T.O. Dill. 1994. "Warm-season Grasses for a Cool Climate." Pp. 206–215 in F.K. Taha, A. Abouguendia, P.R. Horton and T.O. Dill (eds.), *Managing Canadian Rangelands for Sustainability and Profitability.* The First Interprovincial Range Conf. Western Canada, Saskatoon, SK., Jan. 17–20,1993, Grazing and Pasture Technology, Saskatchewan Stockgrowers Association, Regina, SK.

Ward, J.K. 1988. "Optimizing Beef Production Grazing Cool-season and Warm-season Pastures." *Proc. Nebraska Forage and Grassl. Counc. Winter Meeting.*

White, D.J., E. Haber and C. Keddy. 1993. "Invasive Plants of Natural Habitats in Canada: An Integrated Review of Wetland and Upland Species and Legislation Governing Their Control." Report prepared for the CWS, Environment Canada. Cat. No. CW66-127/1993E.

Wilms, W.D., D.J. Major and B.W. Adams. 1994. "The Role of Litter on Forage Production in the Mixed Prairie." Pp.169–174 in F.K. Taha, A. Abouguendia, P.R. Horton and T.O. Dill (eds.), *Managing Canadian Rangelands for Sustainability and Profitability.* The First Interprovincial Range Conf. Western Canada, Saskatoon, SK., Jan. 17–20,1993, Grazing and Pasture Technology, Saskatchewan Stockgrowers Association, Regina, SK.

Wilson, S.D. 2000. "Heterogeneity, Diversity and Scale in Plant Communities." Pp. 53–69 in M.J. Hutchings, E.J. John and A.J.A. Stewart (eds.), *The Ecological Consequences of Environmental Heterogeneity.* Oxford: Blackwell Science.

World Resources Institute. 1990. *World Resources 1990–1: A Report by the World Resources Institute in Collaboration with United Nations Environment Programme and The United Nations Development Programme.* New York: Oxford University Press.

Young, J.A. and C.G. Young. 1986. *Collecting, Processing and Germinating Seeds of Wildland Plants.* Portland, OR: Timber Press.

CHAPTER 8

Abundant Endemic Bugs and Canadian Plains Conservation

K.G.A. Hamilton

Introduction

There is much valid concern about grassland preservation on the changing landscapes of the Canadian plains. Less than 0.05% of the original tall-grass prairies of North America remains unplowed and only about 20% of the more westerly portion is intact (Samson et al., 1998). Where the prairies meet aspen parkland, even less native grassland survives (Ricketts et al., 1999) and prairie boundaries in the eastern half of the Canadian plains become almost wholly conjectural.

With so little native grassland left in Canada, characterizing its original extent and composition is difficult. This is exacerbated by natural changes induced by cyclic environmental conditions, particularly forest/grassland boundaries, which are in turn influenced by human intervention. Wildfires increase the extent of open grassland, destroying aspen and other trees which invade the edges of the prairie. Such fires are generally credited with maintaining the plains as a "fire climax" community (Odum, 1971). This effect may have been amplified by indigenous peoples who set wildfires to replenish pasturage for bison (Mlot, 1990). European settlers since the establishment of the Red River settlement in 1811 suppressed wildfires, allowing trees to regenerate in unproductive grasslands while plowing the most fertile soils for agricultural crops.

As one goes further west, more and more extensive native grasslands are evident (Coupland, 1973). For example, Saskatchewan retains 18.7% (25,000 km^2) of its "mixed grass prairie" (Samson et al., 1998), even greater than its arid rangeland or "short grass prairie" (14.2% or 8,400 km^2). But even the most extensive of these grasslands have become fragmented and altered by human use to such an extent that the character of the original Canadian prairie (Figure 1) as well as its extent and its ecological divisions are debatable. Early concepts of the Canadian prairie (summarized in Scudder, 1979) were hardly less various than recent interpretations (Coupland, 1961; Sims, 1988; Scott, 1995).

The term "prairie" is often used synonymously with "Great Plains." But in the 18th century when La Vérendrye described the grasslands of southern Manitoba (*les prairies* is French for "meadows"), he was painting a vivid word picture of lush, knee-high grassland (Plate 1) similar to water meadows of Europe. These are not at all like the arid lands to the west that were considered in the mid-19th-century as part of the "Great American Desert" (Francis, 1989). Prairies were originally named for the land around Lake Manitoba ("Lac des Prairies") according to annotations on an anonymous 1733 map and again in La Vérendrye's 1737 map

Figure 1. Possibly the only surviving sketch with recognizable plants in authentic tall-grass prairie at a definite site in Manitoba (Dawson, 1875): it represents a view from near Morden looking southwest to the Pembina Hills escarpment. Note that the trees form isolated, fire-pruned clumps and the tall grasses are patchy in distribution, associated with higher ridges of ground and therefore more probably big bluestem and Indian grass rather than cordgrass.

(Combet, 2001). There appears to be little reliable information on the original vegetation. The oldest map specifying dominant land cover was drawn in 1762 by a compiler who used secondary sources (map 48 in Warkentin and Ruggles, 1970); yet the use of the term "Great Meadows" for the land on *both* sides of Lake Manitoba is sufficiently arresting. A more reliable map, drawn in 1859 (map 92 in Warkentin and Ruggles, 1970), specifies that the *entire* western side of Lake Manitoba, from its northern section down to "Lake Petawe Winipeg" (its southern half, south of The Narrows), was "alternate wood and prairie." Today, agriculture has transformed the landscape by expansion of aspen forest with fire suppression and destruction of grassy areas by plowing; this area has taken on the aspect of uniform aspen parkland, except for some land between Lake Manitoba and Lake Winnipeg (the interlake district). South of this area the Red River valley is often designated "tall-grass prairie" (e.g., Scott, 1995) as if comparable to the well-known grasslands of eastern Kansas (Plate 2). Only 0.017% of the deep-soil prairies in Manitoba survive (Joyce, 1989) and most is on the periphery of the Red River valley. The largest preserves are on very rocky, saline, or sandy soil because these were considered of marginal use for farmland. Unfortunately, these very areas are the ones most vulnerable to desertification (Sauchyn et al., this volume).

The Great Plains of Canada differ from those of the US in important ways. They were subjected to less detailed surveys in the 19th century when native vegetation was still extensive. Consequently their original extent and character is more inadequately understood and subject to much more romanticism and overstatement (Francis, 1989). Before railroads and extensive farming came in the 1880s, survey reports of the aridity and fertility of these northern plains differed widely (Lamb, 1977). This was probably due to drought cycles extending for tens of years that have become less evident since then (Sauchyn et al., this volume). Farming is now more intensive in southern Manitoba than in many adjacent parts of the US, while in Alberta there is less use of irrigation and more retention of native grasslands for pasturage.

Reconstructions of Canadian grassland ecosystems are founded on limited data. Early classifications of plains ecosystems were based on climate, principally summer temperature (e.g., Merriam, 1894) and rainfall. Modern classifications of Canadian plains usually follow one of two main criteria: dominant substrate types (Wiken, 1986) or dominant vegetation (e.g., Sims, 1988). Soils in particular are strongly influenced by overall climate patterns such as rainfall and evapotranspiration rates and only to a lesser extent by the kind of vegetation that they support. The areas of "water deficit" and "shortgrass prairie" do not necessarily match. Attempts to harmonize this discordant data hypothesize other biotic boundaries (e.g., Scott, 1995) that are hard to verify or refute due to the highly fragmented nature of remaining native grasslands.

A recent ecological classification citing flora and fauna that are considered to be unique (endemic) to broadly defined "ecoregions" (Ricketts et al., 1999) is also based on areas with ill-defined borders (Hamilton, 2001). But the chief value of ecoregions is the emphasis on endemism, the biogeographic criterion most often cited in conservation efforts. Unfortunately, endemism is most often measured by large species that are highly influenced by changing environmental conditions. What is needed is some measure of endemism based on organisms that are little influenced by drought and fire, survive in a highly fragmented landscape, and yet have enough biodiversity to reveal patterns of wide-ranging significance. Phytophagous insects are abundant enough to be well represented in samples, species-rich on grasslands, and sufficiently studied throughout North America that actual species ranges, microclimate influences, and host associations are known or can be inferred. These criteria are fulfilled best by leafhoppers (family Cicadellidae, Figure 2). Leafhoppers are represented by hundreds of endemic native grassland species, many of which have been intensively collected and studied phylogenetically (Ross, 1970; Whitcomb and Hicks, 1988). These species are usually found only on undisturbed sites on the Great Plains (Hamilton and Whitcomb, 1993) and the prairie-restricted species are almost invariably specialists, feeding only on one or a few closely related native grassland plants. From these insects can be deduced many interesting conclusions about the stability and changes in grassland landscapes and their conservation needs.

Many other insects allied to leafhoppers are found by standard netting and trapping techniques. These include some species endemic to the Great Plains. Cicadas (Cicadidae, Plate 3) are the most familiar of these "true bugs," both for their size and their loud buzzing mating calls. The curiously shaped treehoppers (Membracidae, Figure 3) are locally common. Spittlebugs (Cercopidae, Figure 4) have only one species that is truly a prairie species. Planthoppers (Fulgoroidea, Figures 5–8) are sometimes locally abundant but most species are rarely encountered.

Collectively, all these insects (Homoptera-Auchenorrhycha) are known as "short-horned bugs" because they have unusually tiny antennae with only two basal segments that are easily visible under magnification. They are by far the most species-rich group of animals endemic to the "Temperate Grassland Biome" of Finch and Trewartha (1949). Their endemic species exceed the total of all species of amphibians, birds, butterflies, mammals, reptiles and snails that are endemic to the Great Plains (Ricketts et al., 1999).

Grassland bugs do not seem to be threatened by moderate grazing or by fire, provided that annual burns are avoided (Panzer, 2003). Populations of many species of leafhoppers appear to spread very slowly and cannot invade grassland separated by as little as 16 km (Hamilton, 1999a), yet they can persist on tiny grassy areas even in urban environments. They are therefore ideal candidates for characterizing grassland sites and for mapping areas of faunal similarities.

Short-horned bugs are often widespread in North America but many species are localized. Leafhoppers are particularly likely to be endemic to grasslands, with by

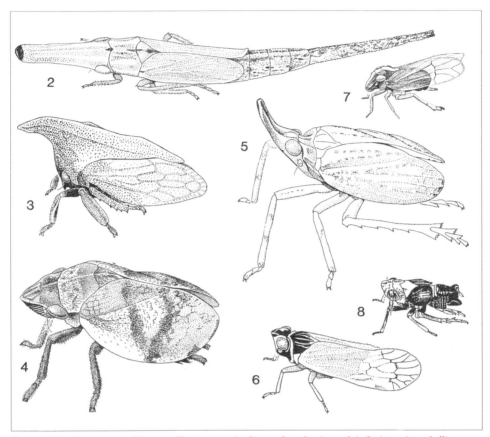

Figures 2–8. "Short-horned" bugs or Homoptera-Auchenorrhyncha (to scale). 2. *Attenuipyga balli*, a completely flightless leafhopper (Cicadelidae); 3. *Campylenchia rugosa*, a common grassland treehopper (Membracidae); 4. *Lepyronia gibbosa*, the only endemic grassland spittlebug (Cercopidae) in Canada, and planthoppers (Fulgoroidea): 5. *Scolops hesperius* (Fulgoridae); 6. *Oliarus dondonius* (Cixiidae); 7. long-winged form of *Bruchomorpha beameri* (Caliscelidae); 8. short-winged form of male *Laccocera flava* (Delphacidae). For Cicadidae, see Plate 3.

far the greatest degree of endemism in semi-deserts of the American southwest (Hamilton, 1999b) and intermontane grasslands of the Pacific Northwest including southern Canada (Hamilton, 2002). Short-horned bugs of grasslands are clearly distinct from those of woodlands. The fauna of the Great Plains has few localized species. Such "regional endemics" are most important in defining native grassland zones, but widespread grassland species that do not simply co-occur throughout the range of their host plants are also important indicators. When they occupy a smaller range than their hosts, as most of them do (*Flexamia*: Whitcomb, and Hicks, 1988), they illuminate the processes by which prairies have evolved and diversified (Whitcomb et al., 1994).

It is the purpose of this analysis to explore the ways in which this rich fauna of grassland-adapted insects can benefit prairie management strategies and to attempt to answer the following questions: Are short-horned bugs in need of conservation? How large must remnant grasslands be to sufficiently retain native insect biodiversity? What characteristics of managed prairies best preserve such biodiversity? Are there significant regional differences in these characteristics? And more vitally, what areas of the Canadian Great Plains have the most of such faunal endemism and are most in need of preservation?

Sampling Short-horned Bugs

Leafhoppers are abundant enough that a sweep net with a canvas bag and about an hour is all that is needed to sample a relatively small or uniform site. A large site can be covered expeditiously if one targets the dominant and subdominant grasses that serve as the main hosts. Knowledge of host species is also important because it assists recognition of host-specific bug species. Despite their small size, many species are easy to recognize (see line drawings in Beirne 1956, 1961; Hamilton, 2000; and paintings in Hamilton, 1985). Common species may be identified visually (if your eyes are sharp and trained) as they scurry or leap up the sides of the net. If a specialist is needed for positive identification the entire catch may be preserved in a jar of alcohol for later sorting. Small specimens can be identified with a hand lens when picked up with a glass-tube aspirator; when using this method, only the rare ones or the ones that are difficult to identify need be killed for later identification.

Leafhoppers are the easiest short-horned bugs to collect. Other short-horned bugs will be found at the same time, so collecting techniques may need to be applied more intensively. Sampling should be done at least twice a year (late May or June, and August) because adults of short-horned bugs live for only a few weeks to a month. If there is time for only a single sampling, a portion of both spring and summer faunas can be found together in mid-to-late July. Continuous sampling by traps will yield the largest number of specimens and successfully find even rare species. Yellow pans or pitfall traps are inexpensive and simple to maintain. Light trapping is acceptable but less efficient because leafhoppers are the only short-horned bugs that come readily to light, and these are very active jumping insects that do not fall passively into the trap. A suction fan or hand picking is required. The light also attracts many other insects such as moths, beetles, and water bugs that will fill the trap quickly if not excluded from the catch.

Early collections of true bugs on the Canadian Great Plains were performed mainly by sweeping. The first intensive work was done by R.W. Salt and J.H. Pepper of the Canadian Department of Agriculture at Lethbridge in the 1930s. Later, A.R. Brooks in Saskatoon collected short-horned bugs in grasslands throughout the prairie provinces for 14 years (1948–62). Herbert H. Ross of the University of Illinois, whose grassland surveys of North American leafhoppers spanned a quarter century (1948–71), made one collecting trip from Saskatchewan to Alaska. Pan trapping at Canadian Forces Base (CFB) Suffield in Alberta and Grasslands National Park in Saskatchewan was carried out in 1995 by Albert Finnamore (Alberta Provincial Museum in Edmonton). Recently, both sweeping and trapping in various sites in southwestern Saskatchewan was done by Jeanette Pepper (University of Regina), and at St. Charles Rifle Range on the western border of Winnipeg by Robert Roughley (University of Manitoba). These studies supplement my own sweep netting over the last 40 years in more than 400 sites across the northern plains, including the northern US.

Sampling Results

More than 600 sites on the Great Plains of Canada have been sampled. Over 340 (the exact number is not known) were found to contain at least one species of

Table 1. Leafhopper genera characteristic of the Great Plains of Canada, with number of grassland-endemic species on the Canadian Great Plains compared to total Canadian fauna.

Acinopterus (1/1)	*Dicyphonia* (1/1)	*Idiocerus* (6/59)	*Paraphlepsius* (7/23)
Aflexia (1/1)	*Driotura* (1/2)	*Laevicephalus* (5/11)	*Pendarus* (1/3)
Amblysellus (5/7)	*Elymana* (1/4)	*Latalus* (1/14)	*Pinumius* (1/1)
Aplanus (1/1)	*Empoasca* (7/94)	*Limotettix* (11/46)	*Polyamia* (1/7)
Athysanella (10/16)	*Erythroneura* (3/95)	*Lonatura* (4/4)	*Prairiana* (2/3)
Attenuipyga (3/5)	*Euscelis* (1/10)	*Macropsis* (3/33)	*Psammotettix* (1/10)
Auridius (4/4)	*Extrusanus* (1/2)	*Macrosteles* (3/30)	*Rosenus* (1/5)
Ballana (1/7)	*Flexamia* (15/16)	*Memnonia* (6/7)	*Stenometopiellus* (1/1)
Ceratagallia (6/17)	*Frigartus* (1/1)	*Mesamia* (3/4)	*Stirellus* (1/1)
Cicadula (1/13)	*Graminella* (1/6)	*Mocuellus* (2/4)	*Stragania* (2/5)
Colladonus (1/28)	*Gypona* (1/2)	*Neohecalus* (1/2)	*Telusus* (1/1)
Commellus (3/5)	*Gyponana* (1/22)	*Neokolla* (1/5)	*Texananus* (4/12)
Cuerna (4/7)	*Hardya* (1/2)	*Norvellina* (2/8)	*Unoka* (1/2)
Deltocephalus (7/14)	*Hebecephalus* (3/10)	*Oncopsis* (2/28)	*Xerophloea* (2/3)
Destria (1/1)	*Hecalus* (2/5)	*Orocastus* (2/4)	

native grassland short-horned bug. Leafhoppers are by far the most common, as 90% of all such records confirm (Hamilton, 2004) and diverse, with 168 grassland species (Table 1). Surprisingly, the second most frequently encountered group (>5% of all records) represents just nine grassland-endemic planthoppers of one small family, Caliscelidae or "piglet bugs" (Table 2). Other planthoppers total 37 Great Plains species in Canada: Delphacidae (31 species), Cixiidae (three species),

Table 2. Other "short-horned" bug genera characteristic of the Great Plains of Canada, with number of grassland-endemic species on the Canadian Great Plains compared to total Canadian fauna.

CICADAS	PLANTHOPPERS	SPITTLEBUGS	TREEHOPPERS
Cicadidae	Caliscelidae	Cercopidae	Membracidae
Okanagana (3/12)	*Aphelonema* (1/1)	*Lepyronia* (1/3)	*Campylenchia* (1/2)
	Bruchomorpha (6/8)		*Ceresa* (1/15)
	Peltonotellus (2/3)		*Leioscyta* (1/1)
			Publilia (1/2)
	Cixiidae		*Telamona* (1/24)
	Cixius (1/15)		*Tortistilus* (1/1)
	Myndus (1/5)		*Vanduzea* (1/2)
	Oliarus (1/11)		
	Delphacidae		
	Caenodelphax (2/6)		
	Criomorphus (1/3)		
	Delphacodes (10/21)		
	Elachodelphax (5/7)		
	Eurybregma (2/3)		
	Laccocera (4/7)		
	Megamelus (1/9)		
	Megamelanus (1/1)		
	Nothodelphax (1/12)		
	Parkana (1/1)		
	Pentagramma (1/3)		
	Pissonotus (1/14)		
	Prokelisia (2/5)		
	Scolopygos (1/1)		
	Yukonodelphax (1/1)		
	Fulgoridae		
	Scolops (1/6)		

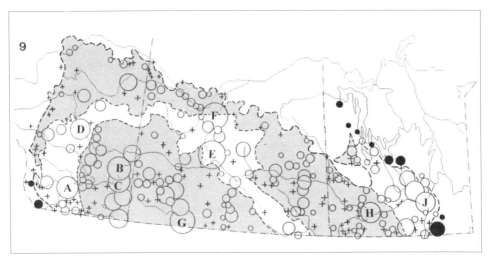

Figure 9. Faunas of "short-horned" bugs on the Canadian Prairies. Circles proportionate to the number of species at each cluster of sites; crosses indicate fewer than four species present. Sites (open circles) superimposed on "aspen parkland" (shaded, upper) and "mixed grassland" ecoregions (shaded, lower); filled circles represent grassland faunas outside of the Prairie Ecozone boundary. A, Lethbridge; B, CFB Suffield; C, Medicine Hat; D, Drumheller; E, Elbow; F, Saskatoon; G, Grasslands National Park; H, Spruce Woods; J, Winnipeg.

and Fulgoridae (one species). There are also a few treehoppers (seven species), cicadas (three species), and a lone spittlebug. Taken all together, more than 2,500 records represent 226 grassland-endemic species.

The number of grassland-endemic species at each site varies tremendously. An average site has 11 grassland-endemic short-horned bug species. Half of the sites have fewer than five such species. By contrast, there are 10 sites where 30 to 50 grassland-endemic short-horned bug species have been found, and four sites (mostly grassland preserves) have more than 60 such species. Grasslands in the vicinity of Elbow, Saskatchewan have no fewer than 86 grassland-endemic species although this area has no grassland preserve.

Most of these extensive faunas were sampled by collectors who were not precise in identifying their specific localities, nor did the collectors keep records of the intensity of collecting. Many such "sites" are probably composites of collections taken at several different sites in the vicinity of a community. Consolidating the faunas of all sites within a 10-km radius should present a less fragmentary picture of a prehistorical Great Plain fauna. The richest of such composite faunas is that of Elbow and vicinity (92 species), or 40% of the entire endemic bug fauna of the Canadian plains. Other major faunas are those of Saskatoon (88) and Grasslands National Park southeast of Val Marie (70) in Saskatchewan; Spruce Woods Forest Reserve (53) and Winnipeg (48) in Manitoba; CFB Suffield (66), Medicine Hat (56), Lethbridge (51) and Drumheller (50) in Alberta. All other composites of plains sites have fewer than 40 grassland-endemic species (Figure 9).

Half of the sites sampled by others are represented by fewer species of grass-land-endemic short-horned bugs than on a patch of ground only 10 m² behind a warehouse in urban Winnipeg. Thus, they probably represent desultory collecting, the debris of sampling that targeted more obvious insects. But even intensive collecting may yield only a few species per site. The reason for this is unclear. Such sites may be very small, in which case their fauna may have degraded by population depletion at the margins, or the few remaining species may simply represent the hardiest species that survived repeated fires or other natural or human-caused disasters.

Small sites should be more subject to population depletion at the margins than large sites. Small sites also offer less botanically and structurally diverse habitats. Nevertheless, one prairie patch at Grosse Isle, Manitoba approximately 95 m long and 30 m wide between the railway right-of-way and roadside is remarkable in having at least 35 bug species restricted to native grasslands. This faunal richness is almost twice that of the largest tall-grass prairie preserves in Minnesota (Hamilton, 1995) and is only slightly less than that of the entire Winnipeg area. Lesser faunas, but otherwise representative for their area, are found on even smaller roadside sites in Manitoba at Morden (25) and Oak Point (20), and in Saskatchewan at Hanley (14) and Stockholm (16). These provide evidence that roadside grasslands in Canada are able to maintain native floral and faunal diversity despite invasive exotic species. Ditches may provide critical buffering zones against changing environmental conditions, as do sand dunes (Hopkins and Running, 2000). In addition, ditches also help protect adjacent prairie from weedy vegetation on roadsides. Native plants are better adapted to fluctuating levels of standing water than species introduced for agricultural or ornamental purposes.

Railway rights-of-way are refugia for prairie plants due to repeated burning during the era of steam engines. Why prairie insect species of such areas should be protected by railway burn-offs in Canada, but not in the US, cannot be explained at present. Perhaps laxer regulations along little-traveled Canadian rail lines did not require annual burn-offs to discourage wildfires, and infrequent burns might promote, rather than decimate, the prairie fauna. A prairie lying between both road and rail corridors is thus doubly protected and important for insect conservation.

Faunal Areas

Short-horned bugs provide abundant evidence to substantiate what ecologists have long recognized (Pitelka, 1941): the biodiversity of the world cannot be neatly subdivided into discrete ecological regions like a jigsaw puzzle. Instead, there are always overlaps (ecotones) between different major community types (biomes) where biodiversity may be higher than in either adjacent area. But this study also highlights the reverse situation: some large areas of the plains appear to have little regional endemism, with only widespread grasslands species reported so far. This complicates but does not negate efforts to classify broad-scale ecological zones.

The Great Plains as a whole appears to be definable by its bug fauna as well as its flora. For example, sites with more than three grassland-endemic species are found commonly throughout the Great Plains in both "prairie" and "aspen parkland" areas (Figure 9, open circles) whereas sites outside the plains usually have much smaller fragments of this fauna. The northernmost such sites correspond closely with the northern boundaries of the plains as defined both by early surveys and by recent soil classifications. The only exceptions occur in Manitoba and southwestern Alberta (Figure 9, filled circles). In the forested foothills of the Rocky Mountains are valleys that contain grassland faunas. Scattered prairie faunas also occur well inside forested areas at the eastern extremity of the plains. This area includes the "tall-grass" preserves of Manitoba around Tolstoi and Gardenton (Plate 4). The easternmost sites agree with an 1874 survey map (117 in Warkentin and Ruggles, 1970) in showing this area to be an extention of the prairies. Similarly, the sites that lie north of the Lake Plains Ecoregion are roughly confirmed as the northern limit of unplowed prairie in a 1945 map (249 in Warkentin and Ruggles, 1970), although the northwestern boundary of that map is merely indicated as hypothetical.

The distribution of grassland leafhoppers shows that the Great Plains are part of a more extensive system of North American grassland faunas. All but 23 of the 55 short-horned bugs that appear to be widespread and characteristic of the Great Plains are intimately associated with smaller native grasslands not connected to the main grassland area (Hamilton, 2004), such as the Peace River district of northwestern Alberta which lies wholly within boreal forest at 56°N latitude. Ephemeral grassland patches, like those on sandy soil in coniferous forests, may have a few of the more widely dispersing short-horned bugs, but in these cases the number of species is very small. Larger faunas are here interpreted as evidence of long-established grasslands. There are at least 18 species of grassland leafhoppers in limestone plains (alvars) of Ontario (Bouchard et al., 2001). There are much larger leafhopper faunas associated with Cordilleran grasslands of the Pacific Northwest (Hamilton, 2002) and some are found as far away as the intermontane valleys of the Yukon (Hamilton, 1997). These are all outliers of the Temperate Grassland Biome embedded in other biomes that are otherwise homogeneously forested. Isolated eastern grasslands have flightless leafhoppers that provide evidence that these were once connected to prairie (Hamilton, 1994).

The four extremities of the Great Plains in Canada have short-horned bug faunas that are strikingly different from each other. Winnipeg represents the easternmost fauna, and Lethbridge the westernmost. Saskatoon represents the most biodiverse northern example of these four types of plains fauna. Of the southernmost sites, Val Marie has the largest number of species found nowhere else. Many of its other species are southern species that barely enter Canada; its fauna differs most strongly from those of the other three faunas. These four faunal suites intermix most obviously at Elbow, Spruce Woods, and Drumheller, which are places of heightened diversity.

Primary subdivisions (biotic provinces) of the Great Plains of Canada are warranted (Figure 10, striped and black half-circles). Each area has more than three dozen characteristic, widespread species of short-horned bugs (Hamilton, 2004). These parts of the plains have leafhopper faunas of equivalent distinctiveness to those of other, isolated parts of the temperate grassland biome including Cordilleran valleys, Great Lakes alvars, and coastal grasslands.

The largest part of the Canadian plains, variously known as the Palliser Triangle or mixed-grass prairie, is a western biotic province that is commonly and more appropriately termed "steppe" (Coupland, 1961). This is a term from Asia for a plain dominated by arid-adapted grasses, including wheat grasses (*Agropyron* spp.) and needle grasses (*Stipa* spp.). In Canada, the steppe has additional "short" grasses such as grama grasses (*Bouteloua* spp.) and June grass (*Koeleria* sp.) that can be locally dominant under favourable conditions. It is maintained by summer drought and cold winters. Wildfires were once common (Francis, 1989) but have been

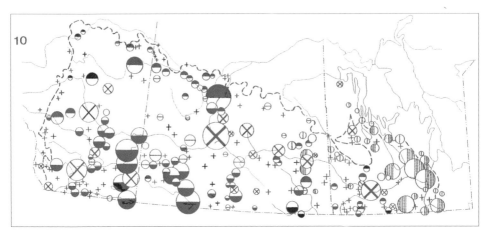

Figure 10. Prairie bug faunas indicating biotic regions: prairie (divided vertically: east side, oak savanna; west side, tallgrass prairie) and steppe (divided horizontally: north side, aspen parkland; south side, short-grass steppe); X mixture; split circle, subdivision of biotic provinces not discernable.

suppressed for more than a century and evidently are not necessary for the persistence of its native flora, albeit in communities altered by human activities.

Steppe extends from isolated sand ridges in southwestern Manitoba to the arid foothills of the Rockies. A total of 34 short-horned bugs are both characteristic and widespread on Canadian steppe, with another 98 species that are more localized (Hamilton, 2004).

The other biotic province is true prairie, characterized by 39 short-horned bug species. This eastern grassland is dominated by grasses of tall to medium height, including bluestem (*Andropogon* spp.), dropseed (*Sporobolus* spp.), and cordgrass (*Spartina* spp.) The western margin of true prairie follows approximately the 400 mm (16") annual precipitation contour. This grassland is maintained by periodic fire, without which it reverts to woodland (Mlot, 1990). After a prairie fire in Illinois, insect populations usually recover within a couple of generations (Panzer, 2003). Such populations appear to peak in Manitoban prairie in the third year after a burn. Following the peak, there is a decline in the numbers of individuals probably due to a build-up of parasites and predators. This suggests that populations adapt to frequent burns, but not to annual ones.

Both biotic provinces, true prairie and steppe, may be subdivided into two biotic regions by their short-horned bug fauna (Figure 10). The subdivisions of the true prairie are here named "tallgrass prairie" and "oak savanna" according to similarities with the faunal and floral composition of such grasslands in the adjacent US. Six bug species are restricted to tallgrass prairie. One of these is a bluestem-specialist, *Flexamia graminea* (DeLong), strongly associated with tallgrass prairie as far south as Texas (Whitcomb and Hicks, 1988). It also ranges into eastern Saskatchewan (Hamilton, in press). Six other species are characteristic of oak savanna: the leafhoppers *Aflexia rubranura* (DeLong) and *Memnonia panzeri* (Hamilton) on dropseed, *Flexamia delongi* (Ross and Cooley), *Paraphlepsius umbrosus* (Sanders and DeLong) and *Polyamia caperata* (Ball) on bluestem, and the planthopper *Bruchomorpha dorsata* (Fitch). A seventh species, *Neohecalus magnificus* Hamilton is also characteristic of oak savanna in Canada although it also invades tallgrass prairie south of Minnesota (Hamilton, 2000), probably along river valleys

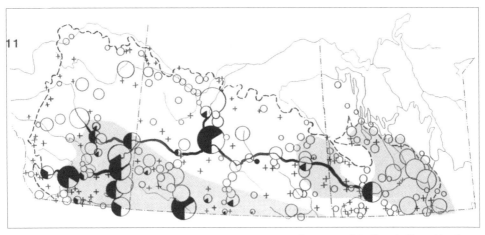

Figure 11. South Saskatchewan-Qu'Appelle-Assiniboine river valley system (black) as indicated by short-grass steppe fauna (shaded, left) and its extension into coulee sites: leafhoppers *Cuerna alpina* and *Laevicephalus exiguus*; planthoppers *Oliarus dondonius, Peltonotellus rugosus*, and *Scolops hesperius;* and tree-hopper *Telamona viridia* (one-sixth of each circle represents one species present; small black dot represents only one species collected at that site). Prairie fauna (shaded, right) also appear to have spread down the same valley system into eastern Saskatchewan.

of the Missouri-Mississippi drainage basin. The dividing line between tallgrass prairie and oak savanna lies along the western edge of the Red River valley and extends north to Lake Manitoba (Figure 10, striped half circles).

Steppe is also subdivided into two biotic regions by its bug fauna (Hamilton, 2004). A northern fauna of 24 characteristic grassland-endemic bug species is found across the aspen parkland of Saskatchewan and Alberta (Figure 9, northern stippled area) but also extends into grassy valleys of the southern foothills of Alberta. Three cicadas are found only in the Rocky Mountain foothills and Cypress Hills of Alberta and these are also associated with aspen groves. This aspen-associated fauna around the perimeter of the plains contrasts with the fauna of the arid central area or "shortgrass steppe." The latter is characterized by nine southern species plus 21 exclusively southwestern species. An additional 34 more widespread western species bring the total shortgrass steppe species to 58. Between aspen parkland and short-grass steppe there is a broad but ill-defined zone (the moist mixed grassland ecoregion) which is an ecotone with a depauperate fauna consisting mainly of widespread species.

Linking all of these biotic regions is a particularly biodiverse east-west corridor that consists of the conjoined glacial meltwater channels that fed Glacial Lake Agassiz and now form the coulees of the South Saskatchewan, Qu'Appelle (Plate 5), and Assiniboine rivers. There, eastern and western species uniquely intermingle. Six western species have migrated down these coulees, three as far as Spruce Woods (Figure 11), the former delta of the Agassiz spillway. More than two dozen southwestern species have expanded their range eastward in the Saskatchewan canyon at least as far as Elbow. Conversely, at least three eastern species have expanded their range west along the Saskatchewan-Qu'Appelle-Assiniboine valley system: the oak savanna leafhopper *Flexamia delongi* and the piglet bug *Bruchomorpha pallidipes* (Stål) are also found in the Qu'Appelle coulee of eastern

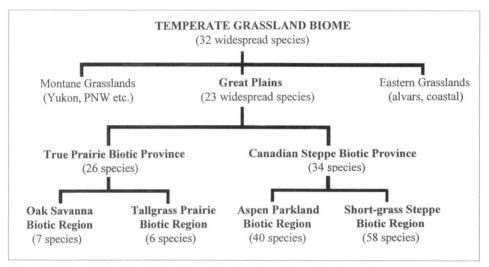

Figure 12. Hierarchal ranking of Canadian plains based on 226 species of its endemic bug fauna, which includes 39 species characteristic of true prairies and 132 of steppe; 55 are common to both biotic provinces. Coulees of the South Saskatchewan-Qu'Appelle-Assiniboine river valleys and hilltops where these faunal suites mix are not included.

Saskatchewan, while the tallgrass prairie leafhopper *Commellus comma* (Van Duzee) is found in the canyon of the south Saskatchewan from Elbow, Saskatchewan to Alberta.

In summary, faunas of endemic short-horned bugs of Canadian grasslands can be arranged in a hierarchal scheme (Figure 12) of diminishing area occupied. These areas in turn appear to correspond to broad zones of ecological similarity that are only weakly indicated by bottom-up hierarchal classifications of community structure, to the point that identifying plant assemblages often becomes more of an art than a science (Radenbaugh, 1998).

The widespread Temperate Grassland Biome that ranges from mountain valleys of British Columbia and the Yukon to Atlantic coastal grasslands is represented by 32 species that occur in both the Great Plains and elsewhere in Canadian grasslands. Another 23 species are restricted to the Great Plains, but occur in both prairie and steppe regions. The plains are subdivided into true prairie, with 26 wide-ranging eastern species, and steppe, with 34 species widespread in Saskatchewan and Alberta. Each of these two biotic provinces are subdivided into two biotic regions. Prairie is subdivided east-west into tallgrass prairie and oak savanna by six and seven species respectively. To the west lies the steppe which is subdivided into peripheral aspen parkland, with 40 distinctive species, and short-grass steppe with 58 grassland-endemic species, separated by a broad ecotonal area. Many of the shortgrass species are highly localized in the Canadian west.

Conclusions

There are hundreds of grassland-endemic short-horned bugs on the Canadian Plains. Dozens of these are either highly localized or at least rare and probably restricted to tiny fragments of native grasslands. Small grassland sites, particularly in the highly endangered tallgrass prairie, are therefore worthy of preservation if they have sufficient topological and soil differences to provide diversified habitats.

Even slight amounts of slope, such as roadside ditches (Plate 6) and railway embankments, can dramatically affect light exposure and moisture gradients, which can significantly increase both floral and faunal diversity. Slopes also help buffer the biota against changing environmental conditions. Where prairie is particularly fragmented, land lying between railway grades and roads (if not burned annually) could serve as biodiversity reservoirs. Extended far enough, railroad prairie could form wildlife corridors connecting refugia, thus increasing the effective extent of native grasslands.

The biodiversity of short-horned bugs on the Canadian plains proves to be neither unique to the region, nor homogeneous. There are many species with isolated populations in remote grasslands, and the most localized endemics are in marginal areas, particularly western valleys. There are also rich areas of regional endemism across wide areas representing eastern and western faunas. These overlap along the length of the Qu'Appelle-Assiniboine river valley system. Where the Qu'Appelle river valley meets the South Saskatchewan River, northern and southern species also overlap resulting in a biodiversity hot spot. This and other sites along the Saskatchewan-Qu'Appelle-Assiniboine system are of previously unsuspected conservation significance.

Acknowledgments

This study is the result on many years of sampling that has been supported and encouraged by the cooperation of numerous individuals and organizations (see Hamilton, 2000: 500). In particular, I am indebted to A.J. Gould, then of Alberta Environmental Protection, Edmonton; D. Nernberg, Last Mountain Wildlife Management Unit, Saskatchewan; and J. Joyce, then of Manitoba Ministry of Natural Resources, Winnipeg, for guiding me to (and around) various sites of special interest. The manuscript has been immensely improved by the comments of C. Bomar, T.A. Radenbaugh, and one anonymous reviewer.

References

Beirne, B.P. 1956. "Leafhoppers (Homoptera: Cicadellidae) of Canada and Alaska," *The Canadian Entomologist* 88, Supplement 2.

Beirne, B.P. 1961. *The Cicadas (Homoptera: Cicadidae) and Treehoppers (Homoptera: Membracidae) of Canada.* Ottawa: Canada Department of Agriculture, Research Branch, Scientific Information Section.

Bouchard, P., K.G.A. Hamilton and T.A. Wheeler. 2001. "Diversity and Conservation Status of Prairie Endemic Auchenorrhyncha (Homoptera) in Alvars of the Great Lakes Region," *Proceedings of the Entomological Society of Ontario* 132: 39–56.

Combet, D. (ed.). 2001. *In Search of the Western Sea: Selected Journals of La Vérendrye.* Winnipeg: Great Plains Publications.

Coupland, R.T. 1961. "A Reconsideration of Grassland Classification in the Northern Great Plains of North America," *Journal of Ecology* 49: 135–167.

——. 1973. "A Theme Study of Natural Grassland in Western Canada: a Report to National and Historic Parks Branch." Canada Department of Indian Affairs and Northern Development, Project 90/7-Pl. Contracts 72-5 and 72-91.

Dawson, G.M. 1875. "On the Superficial Geology of the Central Region of North America." *The Quarterly Journal of the Geological Society of London* 31: 603–623.

Finch, V.C. and G.T. Trewartha. 1949. *Physical Elements of Geography.* New York: McGraw-Hill.

Francis, R.D. 1989. *Images of the West: Changing Perceptions of the Prairies, 1690–1960.* Saskatoon: Western Producer Prairie Books.

Hamilton, K.G.A. 1985. *Leafhoppers of Ornamental and Fruit Trees in Canada* [in English or French]. Ottawa: Agriculture Canada.

——. 1994. "Leafhopper Evidence for Origins of Northeastern Relict Prairies (Insecta: Homoptera: Cicadellidae)." Pp. 61–70 in R.G.Wickett et al. (eds.), *Proceedings of the 13th North American Prairie Conference: Spirit of the Land, our Prairie Legacy.* Windsor: Preney Print & Litho.

——. 1995. "Evaluation of Leafhoppers and Their Relatives (Insecta: Homoptera: Auchenorrhyncha) as Indicators of Prairie Preserve Quality." Pp. 211–226 in D.C. Hartnett (ed.), *Proceedings of the 14th North American Prairie Conference: Prairie Biodiversity*. Manhattan: Kansas State University, Manhattan.

——. 1997. "Leafhoppers (Homoptera: Cicadellidae) of the Yukon: Dispersal and Endemism." Pp. 337–375 in H.V. Danks and J.A. Downes (eds.), *Insects of the Yukon*. Ottawa: Biological Survey of Canada Monograph 2.

——. 1999a. "Leafhoppers (Insecta: Homoptera: Cicadellidae) as Indicators of Endangered Ecosystems." Pp. 103–113 in J. Loo and M. Gorman (compilers), *Protected Areas and the Bottom Line. Proceedings of the 1997 Conference of the Canadian Council on Ecological Areas, September 14–16, 1997*. Fredericton, New Brunswick. Fredericton: Canadian Forest Service.

——. 1999b. "Are Bugs Endangered?" *Proceedings of the 15th North American Prairie Conference*. 104–118.

——. 2000. "Five Genera of New World 'Shovel-headed' and 'Spoon-bill' Leafhoppers (Hemiptera: Cicadellidae: Dorycephalini and Hecalini)," *The Canadian Entomologist* 132: 429–503.

——. 2001. Book review: "Terrestrial Ecoregions of North America: A Conservation Assessment," *Biodiversity* 2, no. 4: 40–43.

——. 2002. "Homoptera (Insecta) in Pacific Northwest Grasslands. Part 2—Pleistocene Refugia and Postglacial Dispersal of Cicadellidae, Delphacidae and Caliscelidae," *Journal of the Entomological Society of British Columbia* 99: 33–80.

——. 2004. "Leafhoppers and Their Relatives (Homoptera, Auchenorrhyncha) from the Canadian Great Plains," *Arthropods of Canadian Grasslands* 10: 16–24.

——. (In press). "Bugs Reveal an Extensive, Long-lost Northern Tall-grass Prairie," *BioScience*.

Hamilton, K.G.A. and R.F. Whitcomb. 1993. "Leafhopper (Insecta: Homoptera: Cicadellidae) Evidence of Pleistocene Prairie Persistence," *American Journal of Botany* 80, no. 6 (supplement): Abstract [from 4th joint meeting of the Botanical Society of America and the Canadian Botanical Association, Ames, IA.] 195: 67.

Hopkins, D.G. and G.L. Running. 2000. "Soils, Dunes and Prairie Vegetation: Lessons from the Sandhills of North Dakota." Pp. 39–57 in T.A. Radenbaugh and P.C. Douaud (eds.), *Changing Prairie Landscapes*. Regina: Canadian Plains Research Center.

Joyce, J. 1989. "Tall-grass Prairie Conservation Project: Final Report." Manitoba Naturalists Society (mimeograph).

Lamb, W.K. 1977. *History of the Canadian Pacific Railway*. New York: Macmillan.

Merriam, C.H. 1894. "Laws of Temperature Control of the Geographic Distribution of Terrestrial Animals and Plants," *National Geographic Magazine* 6: 229–238.

Mlot, C. 1990. "Restoring the Prairie," *BioScience* 40, no. 11: 804–809.

Odum, E.P. 1971. *Fundamentals of Ecology*. Philadelphia: Saunders.

Panzer, R. 2003. "Importance of in situ Survival, Recolonization, and Habitat Gaps in the Postfire Recovery of Fire-sensitive Prairie Insect Species," *Natural Areas Journal* 23, no. 1: 14–21.

Pitelka, F.A. 1941. "Distribution of Birds in Relation to Major Biotic Communities," *American Midland Naturalist* 25: 113–137.

Radenbaugh, T.A. 1998. "Saskatchewan's Prairie Plant Assemblages: A Hierarchal Approach." *Prairie Forum* 23, no. 1: 31–47.

Ricketts, T.H., E. Dinerstein, D.M. Olson, C.J. Loucks, W. Eichbaum, D. DellaSala, K. Kavanagh, P. Hedao, P.T. Hurley, K.M. Carney, R. Abell and S. Walters. 1999. *Terrestrial Ecoregions of North America: A Conservation Assessment*. Washington: Island Press.

Ross, H.H. 1970. "The Ecological History of the Great Plains: Evidence from Grassland Insects." *Pleistocene and Recent Environments of the Central Great Plains. Department of Geology, University of Kansas Special Publication* 3: 225–240.

Samson, F.B., F.L. Knopf and W.R. Ostlie. 1998. "Grasslands." In M.J. Mac, P.A. Opler, C.E. Puckett and P.D. Doran (eds.), *Status and Trends of the Nation's Biological Resources*, Vol. 2. Reston, VA: U.S. Geological Survey, U.S. Department of the Interior.

Scott, G.A.J. 1995. *Canada's Vegetation: A World Perspective*. Montreal: McGill-Queen's University Press.

Scudder, G.G.E. 1979. "Present Patterns in the Fauna and Flora of Canada." Pp. 87–179 in H.V. Danks (ed.), *Canada and its Insect Fauna. Memoirs of the Entomological Society of Canada 108*.

Sims, P.L. 1988. "Grasslands." In M.G. Barbour (ed.), *North American Terrestrial Vegetation*. Cambridge: Cambridge University Press.

Warkentin, J. and R.I. Ruggles. 1970. *Manitoba Historical Atlas: A Selection of Facsimile Maps, Plans and Sketches from 1612 to 1969*. Winnipeg: Historical and Scientific Society of Manitoba.

Whitcomb, R.F. and A.L. Hicks. 1988. "Genus *Flexamia*: New Species, Phylogeny, and Ecology," *Great Basin Naturalist Memoirs* 12: 224–315.

Whitcomb, R.F., A.L. Hicks, H.D. Blocker and D.E. Lynn. 1994. "Biogeography of Leafhopper Specialists of the Shortgrass Prairie: Evidence for the Roles of Phenology and Phylogeny in Determination of Biological Diversity," *American Entomologist* 40, no. 1: 19–35.

Wiken, E.B. 1986. "Terrestrial EcoZones of Canada." Ecological Land Classification Series No. 19. Lands Directorate, Environment Canada. 26pp. and map.

CHAPTER 9

Shifting Sands: Climate Change Impacts on Sand Hills in the Canadian Prairies and Implications for Land Use Management

Stephen Wolfe and Jeffrey Thorpe

Introduction

Sand hills are significant physical features of the prairie provinces, with more than 120 major areas occurring in Alberta, Saskatchewan and Manitoba (Plate 1). These are typically post-glacial sandy deposits that have been reworked into dunes by wind at varying times throughout the Holocene (Muhs and Wolfe, 1999). Surrounded by cultivated lands, they are islands of native grassland or woodland with diverse ecosystems (Hammermeister et al., 2001) and human land uses. In the drier sub-humid prairies, sand hills reside in a delicate balance between bare active dunes and vegetation-stabilized hills that is sensitive to changing climate. Changes in composition and productivity in sand hills vegetation affect their use for livestock grazing, and their role as habitat for native animal species. In northern areas, timber harvesting is also a major land use on forested sand hills. Along the southern margins of the boreal forest, which are presently climatically marginal for tree growth, forests growing on sand hills may be adversely affected by increased aridity.

Climate change has received considerable attention in recent years. While there are still many uncertainties over the degree of change in the near future, there is overwhelming evidence of significant change in the past century. According to the Intergovernmental Panel on Climate Change (IPCC), the average global surface temperature has risen by as much as 0.6°C over the past 100 to 140 years (IPCC WGI, 2001). Most climate researchers relate this change to large anthropogenic increases in "greenhouse gases" such as carbon dioxide (IPCC WGI, 2001). One approach to predicting future climate change relies on complex climate simulations called general circulation models (GCMs). The range of models considered by the most recent IPCC report project further increases in average global surface temperature, ranging from 1.4°C to 5.8°C, by the year 2100. For mid-latitude continental interiors such as the Canadian Prairies, increased summer drying and risk of drought is considered likely as the climate changes (IPCC WGI, 2001).

The sensitivity of sand hills on the Canadian Prairies and their diversity of ecosystems and land uses, make them particularly relevant to investigations of the potential impacts of climate change. A further significance of examining sand hills is that they represent similar landscapes occurring in different settings on the prairies. Therefore, the influences of potential climate change on sand hills within different ecoregions can be compared. Changes in grassland composition and productivity in sand hills in different prairie ecoregions will affect livestock grazing and wildlife habitats to varying extents. In Boreal ecoregions, where forested sand hills occur, timber harvesting may be affected by changes in productivity and vulnerability to insects, fire, and disease. The purpose of this study is to examine the present conditions of sand hills in the prairie provinces and to assess the potential impacts of climate change on the ecology and land use in these areas along with the potential adaptive responses that may be needed to manage these impacts.

Selection of Focus Sand Hills

Within the Canadian Prairie Provinces, there are 76 sand hill areas in Alberta, 43 in Saskatchewan, and six in Manitoba. Data on climate, land use, and land management were collected for all sand hills in order to select a limited set for detailed study. The selected sand hills occur in different present-day climatic conditions and ecoregions. Of particular interest are sand hills in sensitive areas such as the southern Canadian Prairies and those close to ecozone boundaries, such as the Boreal Transition ecoregion. As most of the sand hills are located in the Prairie and Boreal Plains ecozones, a selection of sand hills occurring across these zones was desired.

In selecting the focus areas, sand hills with a range of land use activities were preferred. Land use activities were compared using a matrix arranged by sector. The most common activities include agriculture (occurring on 60% of sand hills) and transportation (on 46%). Recreation/conservation and energy/mines are fairly common (26% and 17%, respectively). Rural and urban communities (10%), forestry (4%), and military land use (4%) are also present in sand hills. Land use activities were unknown for 22% of the 125 sand hill areas, particularly for those located in the northern regions of Saskatchewan and Alberta.

A final consideration in selecting focus areas was the availability of land use management reports. These reports include environmental protection plans, natural resource inventories, environmental impact studies, land use management plans, land use studies, and pasture/range management plans. The selected sand hills required sufficient reports to provide a means of assessing the present sand hills ecosystems and management practices.

Focus Sand Hills

Based on the above criteria, and discussions regarding priority areas with rangeland, park, and forest managers, six focus areas were selected for detailed study (Plate 2). These Canadian sand hills are located in the Mixed Grassland, Moist Mixed Grassland and Aspen Parkland ecoregions of the Prairie ecozone and the Boreal Transition ecoregion of the Boreal Plain ecozone, along a temperature and precipitation gradient moving from warm and dry to cool and wet (Table 1). The area of active dunes in the selected sand hills ranges from large areas to none (Table 1). The land use is diverse (Table 2), with transportation, recreation, and conservation occurring in all six sand hills and grazing occurring in all but one. Detailed management reports are available for most of the selected areas (Table 3).

Middle Sand Hills in southeastern Alberta is a large dune area (328 km^2) in the Mixed Grassland ecoregion of the Prairie ecozone. Source sediments are either alluvial fan deposits or shallow-water lacustrine deposits. The sand hills consist principally of parabolic dunes, with numerous smaller dunes and blowouts active in small areas. The dunes were more active in historical times than at present (Figure 1). The area is encompassed by the portion of the Canadian Forces Base

Table 1. Climatic and ecological and descriptive characteristics of the six focus sand hills areas under current conditions.

SAND HILLS AREA	Middle	Great	Dundurn/ Pike Lake	Manito	Brandon	Fort à la Corne / Nisbet
AREA (km²)	328	1136	312/248	481	964	121/1210
ECOZONE	Prairie	Prairie	Prairie	Prairie	Prairie	Boreal Plain
ECOREGION	Mixed Grassland	Mixed Grassland	Moist Mixed Grassland	Aspen Parkland	Aspen Parkland	Boreal Transition
CLIMATIC NORMALS (1961-90)						
mean annual temperature (°C)	4.7	3.9	2.3	1.9	2.4	0.6
mean June-July-August temperature (°C)	18.6	17.8	17.5	16.3	17.9	16.3
mean Dec-Jan-Feb temp. (°C)	-10.2	-11.1	-15.0	-14.4	-15.4	-17.3
annual precipitation (mm)	300	329	353	413	487	427
potential evapotranspiration (Thornthwaite) (mm)	583	565	551	530	556	517
precipitation / potential evaporation (P/PE)	0.51	0.58	0.64	0.78	0.87	0.83
precipitation – potential evaporation (P-PE) (mm)	-283	-236	-198	-117	-69	-90
OCCURRENCE OF ACTIVE DUNES	small areas	large areas	almost none	small areas	small areas	none

Source: Agriculture and Agri-food Canada (2000).

Suffield that is designated a National Wildlife Area. Because of local environmental sensitivity only limited types of military activity are allowed, and livestock grazing has been terminated. However, oil and gas development are ongoing.

Great Sand Hills in southwestern Saskatchewan is the largest sand dune area (1136 km²) in the Mixed Grassland ecoregion of the Prairie ecozone. Source sediments are glaciofluvial and glaciolacustrine deposits. Individual and compound parabolic dunes with numerous blowouts and elongate ridges are common. The occurrence of active dunes is sporadic, but activity is much higher in the west with large-scale activity in two sections (Plate 3). This area supports a mix of grazing, wildlife habitat, and petroleum land uses. Ecological inventories, land use plans, and petroleum environmental protection plans are all available. The entire area is used for livestock grazing, most of it in large, private ranches (leased crown lands), as well as two provincial community pastures and six grazing cooperatives.

Table 2. Land use by sector in focus sand hills.

LAND USE	Middle	Great	Dundurn/ Pike Lake	Manito	Brandon	Fort à la Corne/ Nisbet
Agriculture – Cultivation			•		•	
Agriculture – Grazing		•	•	•	•	•
Community			•			•
Energy & Mines	•	•		•		•
Forestry						•
Military	•		•		•	
Recreation & Conservation	•	•	•	•	•	
Transportation	•	•	•	•	•	

Table 3. Management reports for land use activities in focus sand hills.

Dune Area	Reports
MB-1 Brandon Sand Hills	Chu et al., 1999 Higgs and Holland. 1999 Manitoba Natural Resources - Parks, 1995 Marr Consulting and Communication, 1995 Schykulski and Moore, 2000
SK-7 Great Sand Hills	Cheeseman, 1997a, 1997b Chivron Environmental Services Inc., 1995 Epp and Townley-Smith, 1980 Lone Pine Resources Ltd., 1990 Saskatchewan Environment and Public Safety, 1991 Thorpe and Godwin, 1997
SK-15 Dundurn Sand Hills	Dillon Consulting, 1998a, 1998b Hilderman Witty Crosby Hanna & Associates et al., 1996 Houston, 2000
SK-23 Manito Lake Sand Hills	Brewster and Schmidt, 1984 Eco-Logic Consulting, 1999 Farrington and Taylor, 1992 Golder Associates, 1997 Manitou Sand Hills Planning and Advisory Committee, 1996 Thorpe and Godwin, 1993 Western Oilfield Environmental Services Ltd., 1997
SK-27 Nisbet Forest Sand Hills	SERM, 2000
SK-28 Fort à la Corne Sand Hills	SERM, 1999
AB-5 Middle Sand Hills	Adams et al., 1997, 1998 Banasch and Barry, 1998 Carbyn et al., 1999 Dale et al., 1999 Didiuk, 1999 Finnamore and Buckle, 1999 Macdonald, 1997 Prairie Farm Rehabilitation Administration, 1998 Reynolds et al., 1999 Shandruk et al., 1998 Wallis and Wershler, 1988

Ongoing oil and gas development is widespread, and the trans-Canada gas pipeline passes through the northern sector of the sand hills. Conservation and dispersed recreation are also important land uses.

Dundurn and Pike Lake Sand Hills, south and west of Saskatoon, Saskatchewan, is the largest sand dune area (444 km² combined) in the Moist Mixed Grassland ecoregion of the Prairie ecozone. Source sediments are glaciolacustrine deposits. The area is covered by blowout hollows, windpits and elongate sand ridges. There is almost no dune activity at present except in disturbed areas, but some stabilized areas were active in historical times (Plate 4). Land use is very diverse, and includes a large military base (Canadian Forces Detachment Dundurn), three PFRA community pastures, a provincial park, municipal conservation areas, private farm and ranch lands, and acreage housing developments associated with the city of Saskatoon. A resource inventory and environmental impact assessment are available for the military base, and the municipal areas are part of a development plan.

Manito Lake Sand Hills in west-central Saskatchewan is a large sand dune area

Figure 1. Active sand dunes in the Middle Sand Hills (Prairies—Mixed Grassland ecoregion) adjacent to the South Saskatchewan River as seen in 1937 (NAPL 5830-50). Dunes shown are presently stabilized.

(481 km²) in the Aspen Parkland ecoregion of the Prairie ecozone. Source materials are sandy glaciolacustrine deposits. The dunes are characterized by elongate parabolic sand dunes, with numerous blowout hollows and windpits. Dune activity is presently confined to the drier south-facing slopes of the elongate sand ridges (Plate 5). All of the available land is used for livestock grazing, with 13 large grazing cooperatives operating on leased crown land. Oil and gas development are ongoing. There are a number of resort lakes in the area, and dispersed recreation is also important. Several management reports are available for this area, including a major land use plan.

Nisbet and Fort à la Corne Sand Hills in central Saskatchewan are large sand dune areas (131 and 1210 km², respectively) in the Boreal Transition ecoregion of the Boreal Plain ecozone. They are located in the rural municipalities of Buckland, Torch River, Nipawin, Willow Creek, and Kinistino. Source materials are sandy glaciolacustrine sediments. The sand hills are comprised of stabilized parabolic dunes containing smaller blowouts. The area is forested and there are no active dunes (Plate 6). The sand hills are on crown land designated as Provincial Forest, with multiple uses including timber-harvesting, wildlife habitat, and recreation. There is also a potential for the development of a large diamond mine. Little management information is available for these areas, but land use plans are being developed.

Brandon Sand Hills is the largest sand dune area (964 km²) in the Aspen Parkland ecoregion. It is also the largest area of sand hills in Manitoba. Source materials are sandy deposits of the Assiniboine delta of glacial Lake Agassiz. The sand hills are comprised of parabolic dunes, blowouts and elongate ridges. They are mostly inactive except in areas deforested for agricultural purposes and a few larger areas such as the Bald Head Hills (Plate 7). The two largest land uses are military training and recreation, including Canadian Forces Base Shilo and Spruce Woods Provincial Park. There is also a PFRA community pasture and private farmland. Some management information is available for Spruce Woods Provincial Park, the military base and the PFRA pasture.

Climate and Ecology in Focus Sand Hills

Five of the six sand hills(Middle, Great, Dundurn/Pike Lake, Manito, and Nisbet/Fort à la Corne) occur along a gradient from warmer/drier to cooler/moister climates across the central part of the prairie provinces (Table 1). Mean annual and seasonal temperatures decline from the Middle Sand Hills to Nisbet/Fort à la Corne, whereas annual precipitation and precipitation-to-potential evapotranspiration (P:PE) ratios increase. The ecoregion classifications reflect this climatic gradient, ranging from Mixed Grassland to Moist Mixed Grassland to Aspen Parkland to Boreal Transition (Table 1). The Brandon Sand Hills (the easternmost dune area) diverges from this gradient in that it has the highest annual precipitation and P:PE ratio, but also has relatively high temperatures, with summers as warm as the Great Sand Hills. The Brandon Sand Hills is a mosaic of grassland and forest, even though it appears to have a moister climate than the completely forested Nisbet/Fort à la Corne Sand Hills, suggesting that our tabulated data may not represent all of the climatic variables needed to account for ecoregion differences.

Limited data are available regarding vegetation productivity in the sand hills (Table 4). Standard grazing capacities from range management publications show a general increase from 0.2 animal unit months (AUM) per acre in the Mixed Grassland to 0.3 in the Moist Mixed Grassland and Aspen Parkland. In the Boreal Transition, this trend is reversed, with low grazing capacities resulting from the high tree cover. Consequently, only Nisbet/Fort à la Corne has significant commercial timber productivity.

Major vegetation types differ along the gradient (Table 4). Sagebrush grassland is a major type only in the two driest areas. Grassland, shrubland (other than sagebrush), and prostrate shrubland (vegetation dominated by creeping shrubs, such as *Juniperus horizontalis*) are found in most areas except the heavily forested Nisbet/Fort à la Corne Sand Hills. Poplar stands are important in all sand hills. Wetlands are a significant part of the dune landscape at the three moistest areas (excluding springs and saline lakes that occur at the margins of sand dunes). Conifer stands are restricted to the two moistest areas, and oak stands (*Quercus macrocarpa*) occur only in the Brandon Sand Hills. Estimates of upland tree cover show a clear increase corresponding to cooler and moister climates across ecoregions.

Climate Change Scenarios and Impact Anaylsis

Scenario Selection

Climate change data from seven internationally recognized models were obtained from the Canadian Climate Impacts Scenarios (CCIS, 2001) project of Environment Canada. For each model, projections are given for four 30-year time periods: 1961–90 (which can be compared to the measured normals), 2010–39, 2040–69, and 2070–99. For convenience, the future time periods are referred to by the middle decade of each period: the 2020s, 2050s, and 2080s. Change values (e.g., difference in temperature or percent change in precipitation) are calculated by comparing model outputs for the future period with those for the base period of 1961–90 (obtained from Agriculture and Agri-food Canada, 2000). Each scenario is based on a "warm start" that begins with the known atmospheric composition of the

Table 4. Productivity and major vegetation types of the six focus sand hills under current conditions.

FOCUS AREA	Middle	Great	Dundurn/ Pike Lake	Manito	Brandon	Fort à la Corne / Nisbet
PRODUCTIVITY						
herbaceous forage yield (lb/ac)		876	775			
grazing capacity (AUM/acre)	0.2	0.20	0.30	0.30	0.30	0–0.15
timber productivity (m3/ha/yr)						1.6–2.8
MAJOR VEGETATION TYPES IN DUNES						
sagebrush grassland	•	•				
grassland	•	•	•	•	•	
shrubland	•	•	•	•	•	
prostrate shrubland		•	•	•	•	
poplar forest	•	•	•	•	•	•
wetland and meadow				•	•	•
oak forest					•	
spruce forest					•	•
pine forest						•
TREE COVER IN UPLANDS	<5%	10%	25%	50%	65%	>95%

Sources
Vegetation:
  Middle Sand Hills: Adams et al., 1997
  Great Sand Hills: Thorpe and Godwin, 1997
  Dundurn Sand / Pike Lake Hills: Houston, 1999
  Manito Sand Hills: Thorpe and Godwin, 1993
  Fort à la Corne / Nisbet Sand Hills: Beckingham et al., 1996; Thorpe, 1990
  Brandon Sand Hills: Chu et al., 1999; Higgs and Holland, 1993
Grazing capacity: Abouguendia, 1990; Wroe et al., 1988
Personal observations and original air photo interpretation of tree cover–J. Thorpe

past century and then runs on the basis of an emissions scenario for the coming century. Some runs include the effects of greenhouse gases only (GG), while others include the effects of both greenhouse gases and sulphate aerosols (GA). Anthropogenic emissions of sulphate aerosols are thought to have a cooling effect on the atmosphere. A few models also provide ensemble-means (GGX or GAX) that are the mean values of several different runs.

The CGCM1 (Canadian model: Boer et al., 2000), ECHAM4 (German model: Roeckner et al., 1996), and HadCM3 (UK model; Pope et al., 2000) were selected for this study, as they cover the entire spectrum of possible changes predicted by the models described above. In general, CGCM1 tends to simulate high temperature changes and moderate to low precipitation levels, ECHAM4 simulates moderate temperatures but covers a wide range of precipitation change, and HadCM3 tends to simulate the least amount of temperature change along with the greatest increases in precipitation. Scenarios based on greenhouse gases only were available for all three models, while greenhouse gases plus aerosols were available for CGCM1 and HadCM3, for a total of five scenarios: CGCM1-GG1, CGCM1-GA1, ECHAM4-GG1, HadCM3-GG1, and HadCM3-GA1.

Scenario Results for Focus Sand Hills

All five models simulated increasing temperatures compared to the present conditions (1961–90 normals) in the focus sand hills. By the 2020s, all models simulate a temperature change of between 1°C and 3°C. Temperature increases on the order of 3°C to 5°C are simulated for the 2050s, with Brandon Sand Hills experiencing the greatest increases. By the 2080s, temperature increase ranges from 3°C to 7°C. The Brandon Sand Hills experience the greatest increases, with temperatures up to 7.5°C warmer than present. Manito and Dundurn/Pike Lake areas experience changes of up to 6.5°C warmer. At all three time steps, CGCM1-GG1 simulates the greatest increases and HadCM3-GA1 the least.

By the 2020s, most models simulate slight precipitation increases (2 mm to 37 mm). Notably, ECHAM4 simulates decreases in precipitation of 2 mm to 42 mm for all dune areas, with the greatest declines in Nisbet/Fort à la Corne. Several other models also simulate precipitation decreases in the Brandon Sand Hills. By the 2050s, almost all models simulate precipitation increases between 1 mm and 51 mm. ECHAM4 simulates only slight changes in either direction, ranging from +7 mm to -20 mm. The wettest scenario is HadCM3-GG1, which simulates an increase of up to 51 mm in Middle, Great, Brandon, and Dundurn/Pike Lake Sand Hills. Overall, Nisbet/Fort à la Corne and Manito Lake Sand Hills experience the greatest average precipitation increases. By the 2080s, most models simulate precipitation increases between 20 mm and 60 mm, with a few extreme cases of >60 mm simulated by CGCM1-GA1 and HadCM3-GG1. In contrast, ECHAM4-GG1 simulates decreases between 0 mm and 36 mm in most areas, with a few grid cells experiencing slight increases of up to 20 mm. As with the 2050s, Manito Lake Sand Hills appear to experience the greatest precipitation increases.

A detailed analysis carried out for the 2050s shows annual surface moisture availability through P:PE ratios and precipitation surpluses/deficits based on the Thornthwaite model (Thornthwaite and Mather, 1970) of potential evapotranspiration (Tables 5 and 6). Almost all scenarios and all areas show a decrease in P:PE, indicating a shift towards lower surface moisture availability (Table 5). This decrease tends to be more pronounced in moister areas and less so in drier areas, with the Great and Middle Sand Hills showing no change under one scenario. Only the Middle and Manito Lake Sand Hills shift climatic classifications (i.e. dry sub-humid to semi-arid, and humid to sub-humid respectively). Brandon and Nisbet/ Fort à la Corne Sand Hills shift from humid to sub-humid using the warmer (CGCM1-GG1) and drier (ECHAM4) scenarios, but remain unchanged for the cooler and wetter scenarios (HadCM3, CGCM-GA1). Although Dundurn/Pike Lake and Great Sand Hills largely remain in the dry sub-humid classification, they both show decreased P:PE ratios and are considerably closer to the semi-arid threshold.

Normals for the dune areas show that they currently experience a precipitation deficit between 66 mm and 283 mm (Table 6). All areas show a greater deficit under the various climate change scenarios, although the size of the change varies. These precipitation deficits are greatest in the CGCM1 and ECHAM4 simulations.

Potential Impact Analysis

Potential impacts of climate change on the vegetation in the six focus areas were analyzed in three ways. The first was by using present-day analogues for simulated future climates. Areas with a warmer climate similar to that projected for

Table 5. Ratio of annual precipitation to potential evapotranspiration (P:PE) in the focus sand hills areas: observed 1961–90 normals, and scenario outputs for the 2050s.

Dune Area Name	1961–90 Normals	CGCM1		ECHAM4 GG1	HadCM3	
		GG1	GA1		GG1	GA1
Brandon Sand Hills	0.88	0.74	0.76	0.75	0.84	0.79
	0.87		0.74	0.71	0.82	0.78
Fort à la Corne / Nisbet Sand Hills	0.86	0.74	0.81	0.74	0.81	0.81
	0.80	0.69	0.75	0.66	0.74	0.72
Manito Lake Sand Hills	0.78	0.69	0.72	0.68	0.74	0.73
		0.68		0.66	0.73	0.71
Dundurn/Pike Lake Sand Hills	0.64	0.58	0.60	0.55	0.63	0.59
		0.56	0.55		0.60	0.58
Great Sand Hills	0.58	0.52	0.51	0.49	0.58	0.53
				0.46		
Middle Sand Hills	0.51	0.46	0.49	0.43	0.52	0.47
		0.44	0.45	0.40	0.50	0.46

Aridity Index or P:PE ratio

0.76-1.00 Humid	0.66-0.75 Sub-humid	0.51-0.65 Dry sub-humid	0.2-0.5 Semi-arid

Table 6. Difference between annual precipitation and potential evapotranspiration (P-PE) in the focus sand hills areas: observed 1961–90 normals, and scenario outputs for the 2050s.

Dune Area Name	1961–90 Normals (mm)	CGCM1		ECHAM4 GG1 (mm)	HadCM3	
		GG1 (mm)	GA1 (mm)		GG1 (mm)	GA1 (mm)
Brandon Sand Hills	-73	-180	-174	-193	-115	-141
	-66	-175	-160	-165	-104	-134
Fort à la Corne / Nisbet Sand Hills	-106	-188	-145	-208	-153	-167
	-75	-158	-113	-155	-110	-110
Manito Lake Sand Hills	-117	-206	-172	-212	-161	-175
		-198	-171	-193	-157	-164
Dundurn/Pike Lake Sand Hills	-198	-297	-295	-295	-253	-265
		-276	-244		-233	-254
Great Sand Hills	-236	-331	-324	-368	-257	-299
				-337		
Middle Sand Hills	-283	-395	-376	-428	-333	-356
		-384	-318	-393	-323	-354

Precipitation Index

> -100 mm	-100 to -200 mm	-200 to -300 mm	< -300 mm

the focus areas can presently be found to the south in the US. These analogue areas illustrate the kind of vegetation that could be supported in the focus areas by a warmer climate. However, we emphasize that we are not predicting that this kind of vegetation will occur in the focus areas under the warmer climate. The vegetation presently established has considerable inertia, and significant change may depend on specific disturbances (e.g., fire and grazing) that remove the current vegetation. Moreover, new species may vary in the speed with which they expand their ranges northward, and some may be impeded by the current fragmentation of native habitats. Therefore, analogue analysis should be seen as showing a potential direction of change, rather than a definite prediction of future composition.

Analogue analysis was based on the climate change scenarios for the 2050s. Future values of temperature and precipitation from these scenarios were compared to climatic patterns in the US Great Plains to select the approximate regions having analogous climates. Climatic data for each dune area were obtained by selecting several climate stations in the vicinity, taking the general averages of the 1961–1990 normals (NOAA, 2004).

The main analogue source was the Nebraska Sand Hills, a region of about 5 million ha (Adams et al., 1998), as well as smaller areas of dunes extending from Nebraska into northeastern Colorado (Ramaley, 1939) and southern South Dakota. There is extensive scientific literature related to the vegetation of the Nebraska Sand Hills (Ramaley, 1939; Weaver, 1965; Wolfe, 1972; Bragg, 1978; Adams et al., 1998; Bleed and Flowerday, 1989). However, the best source of information for vegetation in relation to climate is the US Natural Resources Conservation Service (NRCS) range site descriptions for the "sands" and "choppy sands" range sites, which were provided by NRCS in Nebraska. These descriptions are divided according to a series of precipitation zones from east to west; the two westernmost of these, the 14–16 inch (355–405 mm) and 17–19 inch (430–80 mm) zones, were also applied to similar climates in northeastern Colorado. Analogue information was also found for somewhat cooler climates in central North Dakota, and vegetation descriptions were again obtained from range site descriptions (provided by NRCS staff in North Dakota) for the "sands" and "thin sands" range sites.

A second means of assessing the potential impact on sand hills was examination of the relations between climate and the productivity of grasslands using simple regression models. Most studies from the US Great Plains have focused on changes in productivity with precipitation as this is the prominent east-west gradient in the region (Sims et al., 1978; Sala et al., 1988; Schimel et al., 1990). However, two models are available for examining effects of both precipitation and temperature on productivity. Sims et al. (1978) analyze data from ten sites across the US Great Plains studied under the International Biological Program. Epstein et al. (1997) also generate regression equations for the percentages of C3 (cool-season) and C4 (warm-season) plants in the community. These production models were applied to climatic data for the focus areas, and the percentage change in plant communities from the 1961–90 base period to the 2050s was calculated.

Lastly, the change in dune activity in the focus dune areas was predicted from the model applied by Muhs and Maat (1993) and Wolfe (1997), incorporating wind strengths and the ratio of annual precipitation to potential evapotranspiration (P:PE). The potential changes in dune activity were assessed based on 2050s changes in P:PE ratios, assuming no change in wind regime.

Potential Impacts of Climate Change on Sand Hills

Analogue Analysis

The climatic relations between the focus study areas and the analogue areas were examined by plotting mean annual, winter and summer temperatures against annual precipitation (Thorpe et al., 2001). The relation between summer temperature and annual precipitation is shown in Plate 8. For assessing impacts on vegetation, summer temperatures are the most relevant as they relate to the amount of heat available in the growing season, which typically shows a close relation to vegetation zonation (Tuhkanen, 1980). The distribution of the single points representing 1961–90 normals shows the climatic gradient among the six focus areas, as discussed earlier. The 2050s projections (the ellipse enclosing five points for each focus area) show that there is some variability among the climate change scenarios. In spite of this variability, trends in future change are apparent. For the 2050s, all of the scenarios show increases in temperature, and most of them show increases in precipitation.

The 2050s climates of the focus areas approach the current climates of analogue areas in the US (Plate 8). The lower part of the graph shows the three driest study areas (Middle, Great, and Dundurn/Pike Lake Sand Hills) approximating the climate of the 14–16 inch (355–405 mm) precipitation zone in western Nebraska and northeastern Colorado. Actually, the Middle Sand Hills, and to a lesser extent the Great Sand Hills, are projected to be even warmer and drier than the current climate of this analogue region. However, examination of US climate records shows that these warmer and drier climates are found only a little further south (in central Colorado) and there is no indication in the literature of an associated change in dune vegetation. The coolest areas (Manito Lake and Nisbet/Fort à la Corne) of the 2050s are similar to the present climate of central North Dakota. The simulated future climate of the Brandon Sand Hills approximates that of the present 20–24 inch (510–610 mm) zone in the Nebraska Sand Hills.

A significant finding from this analogue approach is that all of the focus areas could experience climatic conditions that are less suitable for tree growth. This impact will vary among focus areas, which currently differ from low to high tree cover along the gradient from dry to moist climates. The impact could be greatest for the Nisbet and Fort à la Corne Sand Hills, currently almost entirely forested, and they could shift to a climate like central North Dakota that supports only grassland (Thorpe et al., 2001: Table 14). The shift from a mosaic of woody and grassy types towards more uniform grassland will have a significant impact on all of the areas.

Summary of Potential Changes in Dune Areas

Detailed analytical results of potential changes in dune activity, vegetation zonation, species composition, and grassland productivity are provided in Thorpe et al. (2001). Herein, we will report a partial summary of the impact analysis.

Because of the shift to effectively drier climates due to lower P:PE ratios, the potential for dune activity will increase, particularly in the driest areas (Dundurn/Pike Lake, Great, and Middle Sand Hills). This finding is similar to results that were obtained by Wolfe (1997). Dune activation is not a uniform process, but may develop in certain locations following a series of dry years. If significant areas of active dunes develop, this could reduce the amount of productive vegetation for livestock grazing and wildlife habitat. This increased potential for dune activity could also increase erosion problems associated with roads, wellsites, or military operations.

In grasslands, the climate could increasingly favour warm-season (C_4) over cool-season (C_3) species (Thorpe et al., 2001: Figs. 15 and 16), though the carbon fertilization effect of increased atmospheric CO_2 will have additional effects on grassland composition and productivity (Ferretti et al., 2003) that are not accounted for herein. Native warm-season grasses, including sand grass (*Calamovilfa longifolia*), sand dropseed (*Sporobolus cryptandrus*), blue grama (*Bouteloua gracilis*), and, at Brandon, big bluestem (*Andropogon gerardii*), presently grow in the sand hills. Thus, grasslands will have a considerable capacity to adjust to climate change through shifts in the relative proportions of species that are already present. Areas will probably also experience a gradual northward migration of species currently absent or uncommon in Canadian grasslands, especially warm-season grasses such as sand bluestem (*Andropogon hallii*), sand muhly (*Muhlenbergia pungens*), and switchgrass (*Panicum virgatum*).

Shifts in vegetation patterns may be accompanied by shifts in the populations and distributions of other species, because of habitat changes and direct climatic effects (e.g., temperature tolerances or amount of snow cover). A loss of forest cover would also result in loss of forest-dependent plant and animal species. Furthermore, habitats may become less suitable for "edge" species such as white-tailed deer (*Odocoileus virgianus*), but more suitable for open grassland species. On the other hand, any tendency towards dune activation may help to maintain habitats for plant, mammal, and arthropod species—for example, lance-leaved psoralea (*Psoralea lanceolata*), Ord's kangaroo rat (*Dipodomys ordii*), tiger beetles (*Cincindela* spp.)—that require active dunes. New dune-adapted species from more southerly ranges in the US Great Plains may gradually migrate northward. Disruption of local plant communities could increase opportunities for invasive Eurasian exotic species.

The impact on grazing capacity is unclear. A shift to a somewhat drier climate suggests reduced forage production, and this is supported by simple production models which show decreased yields in the 2050s (Thorpe et al., 2001: Fig. 14). However, the current grazing capacities of analogue areas in the US are rated higher than in the Canadian focus areas, with the higher proportion of warm-season grasses and the lower woody cover possibly contributing to this difference. Forested areas such as Nisbet/Fort à la Corne are likely to increase in grazing capacity with loss of tree cover.

Adaptive Responses of Land Use Activities to Climate Change

If climate does change as projected by current GCMs, many physical, biological, economic and social impacts can be expected as outlined in Herrington et al. (1997). In order to minimize the harmful effects of these changes, and to maximize the potential benefits, it is important to consider human adaptation options and to develop strategies for coping with climatic change.

Because most of the focus sand hills areas do not have formal management plans, a number of documents related to management planning (including environmental assessments, wildlife inventories, and research reports) were evaluated (Table 3). In addition, a stakeholder workshop provided first-hand insight into land uses, issues, and possible adaptation options in the focus areas. Many land uses are not based on formal management plans. However, they may still be considered intentional management actions that can be revised in response to climate change. An assessment by sector of potential adaptive responses to climate change is presented in the following sections.

Livestock Grazing

Within most sand hills on the prairies, recommended stocking rates and range condition assessment tables should be revised (Wroe et al., 1988; Abouguendia, 1990) as new information on changes emerges from decadal-scale monitoring of range benchmark sites. The benefits and ecological consequences of introducing warm-season grasses such as sand bluestem from nearby North American ranges could be reviewed. While reseeding of existing grasslands is undesirable, use of such species in reclamation may be a viable adaptation to climate change. It is likely that better fire-fighting capabilities will be required to protect rangelands from an increased fire hazard.

In northern sand hills such as Nisbet/Fort à la Corne, livestock grazing may increase in importance, perhaps becoming the dominant land use as forests decline. Stocking rates used as a basis for issuing Provincial Forest grazing permits may have to be adjusted to reflect vegetation changes observed through monitoring.

Forestry

Issues pertaining to forestry in this study are restricted to the Nisbet and Fort à la Corne Sand Hills (see SERM, 1999 and 2000), but they also may be relevant across much of the Boreal Transition ecoregion where commercial forestry currently occurs. Calculations of allowable harvest, and harvest allocations to individual operators, may require adjustment if evidence from monitoring programs indicates changes in growth rate or incidence of disturbance. Increased spending on regeneration following harvest or disturbance may be necessary to cope with regeneration failures. Research on optimal harvest and regeneration methods for dry conditions may be required. Increased fire protection may be needed as frequency of high fire hazard increases. Planned timber harvesting to interrupt the continuity of high-hazard fuels may need to be implemented. The benefits and ecological consequences of introducing species from nearby North American ranges, such as red pine (*Pinus resinosa*; native in southeastern Manitoba), could be reviewed. Use of such species in regeneration programs may help to maintain forest cover in a warmer climate and address the dwarf mistletoe problem in the native jack pine (*Pinus banksiana*). Timber-using industries should plan for the possibility of reduced timber supply from climatically marginal areas such as the Nisbet/Fort à la Corne Sand Hills.

Oil, Gas, and Mining Development

Policies and regulations are presently in place to control the impact of oil/gas activities on soil erosion and other environmental issues (see Golder Associates, 1997; Lone Pine Resources Ltd., 1990; Saskatchewan Environment and Public Safety, 1991). However, it may be necessary to apply current policies more stringently as the climatic potential for dune activation increases. The area of terrain considered too sensitive for oil/gas development may expand, thus increasing the impetus for development of low-impact practices for drilling and extraction and for more vigorous reclamation. The option of winter-only operations (e.g., for well-drilling) may become less viable in southern areas as milder winters reduce the period during which soil is frozen.

Military

An increased potential for dune activation should require greater limitations on military activities on sensitive sites and increased reclamation efforts where erosion has started. With increasing frequency of high fire hazard, more resources will be required for fire prevention and control.

Recreation

In most sand hills areas the existing dispersed recreation patterns will probably require minimal adaptative adjustment. However, sites sustaining intensive recreation activity, such as in the Dundurn/Pike Lake and Brandon (see Manitoba Natural Resources Parks, 1995) Sand Hills may require additional trail development to control pedestrian impacts and more reclamation of disturbed sites. In the Brandon Sand Hills, the need for active management to control brush encroachment on grassland patches (Schykulski and Moore, 2000) could be eliminated as the climate shifts to one less conducive to woody growth. Monitoring will also be needed to detect any change in brush encroachment in the coming decades. In the Nisbet and Fort à la Corne Sand Hills, forest-based recreation facilities such as snowmobile and ski trails (SERM, 1999 and 2000) may require rerouting. In general, greater regulation may be required for motorized recreation vehicle use on sensitive sites.

Conservation

All focus areas are to some extent remnant islands of natural vegetation, and as such are considered important for conservation of wildlife habitat and biodiversity. Current concepts of conservation are based on maintaining all of the species and ecosystems that are currently present (or were present in recent history). However, climate change could result in shifts in current patterns as some species would be less able to adapt to new conditions. In such a situation, a rethinking of conservation policy would be required. Should local decline in a particular species be opposed or accepted as inevitable under the changed climate? This question is particularly relevant to the management of areas used primarily for conservation, such as Suffield National Wildlife Area and Spruce Woods Provincial Park (see Table 3 references). The consequences of intentional introductions of species from warmer parts of North America also require more study. Under current conservation thinking, such introductions into areas of natural vegetation are opposed because these species are exotic to the local area. However, species that are native to nearby areas may eventually arrive anyway, and are less likely than Eurasian exotics to be ecologically disruptive. Their intentional introduction at an early stage may be a useful strategy for adapting to climate change. Research into such introductions and their ecological consequences will aid in the development of policy regarding their use as an adaptation option. Any increase in opportunities for invasion by Eurasian exotics will reinforce the need for cautionary measures such as screening intentional introductions for invasive potential, preventing unintentional transport (e.g., in contaminated seed), avoiding creation of establishment sites (e.g., disturbed roadsides or well-sites), and active control of existing populations.

Summary

On the basis of the scenarios analyzed in this study, the projected climate change in focus areas on the Canadian Prairies by the 2050s may include: an increase in average temperatures in all areas, with changes on the order of 2.5 to 5.0°C, and warming will occur in all seasons; a small increase in annual precipitation in all areas (scenarios vary from essentially no change to an increase of about 50 mm); and a decrease in moisture availability in all areas. P:PE ratios will decrease more markedly in the moister areas and less so in the drier ones.

Under these projected changes, there are likely to be significant influences to sand dune areas in the Canadian Prairies. These impacts may include changes in grassland species composition and productivity, reduction in forest productivity, and potential increases in the susceptibility of sand hills to erosion. These physical and biological impacts, in turn, may have social and economic consequences that will require society to alter its behaviour and respond with adaptive strategies that minimize negative economic and environmental consequences with the climate change.

Management planning and practices in most sand hills areas are clearly more protective and conservation-oriented than in other parts of the prairies. Progress on land-use planning is also comparatively advanced (especially in Great and Manito Lake Sand Hills). This situation should facilitate adaptation to climate change. However, detailed biophysical inventories should be completed for all areas to obtain a baseline for monitoring. Land use plans should be developed for all areas and should consider the impacts of climate change on future land uses. Land use plans should incorporate mechanisms for adaptive management (e.g., monitoring, scheduled plan review, and revision). Communication with managers and other stakeholders should continue to increase awareness of climate change impacts and adaptation. Vegetation monitoring programs should be designed to include indicators of climate change impacts, such as decadal-scale change in grassland productivity and species composition measured in a network of fenced range benchmark sites. Repeat time-series aerial photography or satellite imagery can be used for monitoring of changes in the woodland/grassland mosaic over large areas.

As the Boreal Transition ecozone is susceptible to change, forest monitoring programs in areas such as the Nisbet and Fort à la Corne Sand Hills should be designed to include indicators of climate change impacts. Such indicators may include changes in tree growth rates, drought-related tree mortality, incidence of insects and diseases—for example, forest tent caterpillar (*Malacosoma disstria*), spruce budworm (*Choristoneura fumiferana*), dwarf mistletoe (*Arceuthobium americanum*)—and changes in plantation survival.

Acknowledgments

This study was carried out through funding provided by the Climate Change Action Fund (CCAF) and the Prairie Adaptation Research Collaborative (PARC) to projects QS-11 and PS-07. Funding and in-kind contributions were also provided by the Geological Survey of Canada and the Saskatchewan Research Council. The authors wish to acknowledge the assistance of Janet Campbell, Jennifer Leblanc, and Rachel Molder for their efforts in this project. Additional assistance was provided by Joe Park, Zoe Pfeiffer, and Mark Smith at the Geological Survey of Canada, and Charlene Hudym and Leanne Crone at Saskatchewan

Research Council. The authors are also indebted to many collaborators who assisted with obtaining management documentation and who participated in the workshop. In particular, we wish to thank Andy Didiuk, Bill Houston, Harry Loonen, Lori Nichols, Del Philips, Beatrice Regnier, Jim Schmidt, and Lorne Veitch for presentations at the workshop and Eleanor Bowie, Wayne Harris, and Bill Houston for their efforts throughout the project. The manuscript benefited from review by Dan Muhs and Elmo Rawling and editing by Glenn Sutter and Todd Radenbaugh. This article represents a contribution to the Reducing Canada's Vulnerability to Climate Change program, GSC contribution number 2003101.

References

Abouguendia, Z.M. 1990. "A Practical Guide to Planning for Management and Improvement of Saskatchewan Rangeland—Range Plan Development." New Pasture and Grazing Technology Project. Saskatchewan Agriculture Development Fund: Regina, Saskatchewan.

Adams, D.C., R.T. Clark, P.E. Reece and J.D. Volesky. 1998. "Research and Education for Managing Resources Within the Nebraska Sandhills: The Gudmundsen Sandhills Laboratory," *Rangelands* 20: 4–8.

Adams, G.D., G.C. Trottier, W.L. Strong, I.D. MacDonald, S.J. Barry, P.G. Gregoire, G.W. Babish and G. Weiss. 1997. *Vegetation Component Report—Canadian Forces Base Suffield National Wildlife Area, Wildlife Inventory*. Edmonton: Canadian Wildlife Service, Environment Canada.

Agriculture and Agri-Food Canada (AAFC). 2000. Canadian Ecodistrict Climate Normals 1961–1990. http://sis.agr.ca/CANSIS/NSDB/ECOSTRAT/DISTRICT/climate.html. (Last updated February 16, 2000. Accessed November 17, 2000).

Banasch, U. and S.J. Barry. 1998. *Raptor Component Report—Canadian Forces Base Suffield National Wildlife Area, Wildlife Inventory*. Edmonton: Environment Canada, Prairie and Northern Region.

Beckingham, J.D., D.G. Nielsen and V.A. Futoransky. 1996. *Field Guide to the Ecosites of Mid-Boreal Ecoregions of Saskatchewan*. Canadian Forest Service, Northern Forestry Centre, Special Report 6.

Bleed, A. and C. Flowerday. 1989. *An Atlas of the Sand Hills (Resource Atlas No. 5a)*. Lincoln: Conservation and Survey Division, Institute of Agriculture and Natural Resources, University of Nebraska.

Boer, G.J., G. Flato and D. Ramsden. 2000. "A Transient Climate Change Simulation with Greenhouse Gas and Aerosol Forcing: Projected Climate to the 21st Century." *Climate Dynamics* 16: 427–450.

Bragg, T.B. 1978. "Effects of Burning, Cattle Grazing, and Topography on Vegetation of the Choppy Sands Range Site in the Nebraska Sandhills Prairie." In D.N. Hyder (ed.), *Proceedings of the First International Rangeland Congress, August 14–18*. Denver: Society for Range Management.

Brewster, D. and J. Schmidt. 1984. "The Manito Sandhills: A Wildlife-Agriculture Development Proposal." Saskatchewan Parks and Renewable Resources and Saskatchewan Agriculture, Joint Technical Report 84-1.

Canadian Climate Impacts and Scenarios (CCIS). 2001. Web-site. http://www.cics.uvic.ca/scenarios/index.cgi/Introduction (Last updated April 24, 2003. Accessed November 15, 2000).

Carbyn, L.N., M. Woelfl and H. Schinke. 1999. *Carnivore Component Report—Canadian Forces Base Suffield National Wildlife Area, Wildlife Inventory*. Edmonton: Canadian Wildlife Service, Environment Canada.

Cheeseman T. 1997a. Great Sand Hills Planning District: Development Plan, Bylaw No. 1-98. Rural Municipality of Clinworth.

———. 1997b. Great Sand Hills Rural Municipality Clinworth No. 230: Zoning Bylaw No. 2-98. Rural Municipality of Clinworth.

Chivron Environmental Services Inc. 1995. Environmental Impact Statement 1994/1995 Freefight. Ocelot Energy Inc. Prepared for Ocelot Energy Inc., Project No. 3189.

Chu, G., W. Lenfesty, and J. Birnie. 1999. *Range Management Plans for Langford Community Pasture*. Regina: Agriculture and Agri-food Canada, Prairie Farm Rehabilitation Administration

Dale, B.C., P.S. Taylor and J.P. Goossen. 1999. *Avifauna Component Report—Canadian Forces Base Suffield National Wildlife Area, Wildlife Inventory*. Edmonton: Canadian Wildlife Service, Environment Canada.

Didiuk, A.B. 1999. *Reptile and Amphibian Component Report—Canadian Forces Base Suffield National Wildlife Area, Wildlife Inventory*. Edmonton: Canadian Wildlife Service, Environment Canada.

Dillon Consulting Ltd. 1998a. "Canadian Forces Detachment Dundurn Natural Resources Inventory (draft report)." Department of Natural Defense, 96-3604-04-01.

——. 1998b. "Canadian Forces Detachment Dundurn Environmental Assessment (draft report)." Defense Construction Canada, 96-3604.

Eco-Logic Consulting. 1999. "Environmental Protection Plan for the Wascana Energy Inc. 1999/2000 Winter Main Heavy Oil Horizontal Project." Prepared for Saskatchewan Environment and Resource Management, Regina, Saskatchewan.

Epp, H.T. and L. Townley-Smith. 1980. *The Great Sand Hills of Saskatchewan.* Regina: Saskatchewan Environment, Policy and Research Branch.

Epstein, H.E., W.K. Lauenroth, I.C. Burke and D.P. Coffin. 1997. "Productivity Patterns of C3 and C4 Functional Types in the U.S. Great Plains." *Ecology* 78: 722–731.

Farrington, P.D. and P.S. Taylor. 1992. "Significant Migratory Bird Habitats and Species of the Manitou Sand Hills Study Area." Unpublished Report. Canadian Wildlife Service. Saskatoon, Saskatchewan.

Ferretti, D.F., E. Pendell, J.A. Morgan, J.A. Nelson, D. LeCain and A.R. Mosier. 2003. "Partitioning Evapotranspiration Fluxes from a Colorado Grassland Using Stable Isotopes: Seasonal Variations and Ecosystem Implications of Elevated CO2." *Plant and Soil* 254: 291–303.

Finnamore, A.T. and D. Buckle. 1999. *Arthropod Component Report: the Stinging Wasps (Hymenoptera: Chrysidoidea, Vespoidea, Apoidea) and Spiders (Araneae)—Canadian Forces Base Suffield National Wildlife Area, Wildlife Inventory.* Edmonton: Canadian Wildlife Service, Environment Canada.

Golder Associates. 1997. "Environmental Protection Plan for the Proposed Winter Oil Recovery Project 8-42-25 W3M." Prepared for Gulf Canada Resources Limited.

Hammermeister, A.M., D. Gauthier and K. McGovern. 2001. *Saskatchewan's Native Prairie: Statistics of a Vanishing Ecosystem and Dwindling Resource.* Saskatoon: Native Plant Society of Saskatchewan Inc.

Herrington, R., B. Johnson and F. Hunter. 1997. *Responding to Global Climate Change in the Prairies.* Volume III of the Canada Country Study: Climate Impacts and Adaptation. Adaptation and Impacts Section, Atmospheric Environment Branch, Environment Canada, Prairie and Northern Region.

Higgs, C.D. and G. Holland. 1999. "Natural Resource Inventory of Spruce Woods Provincial Park, Manitoba." Manitoba Natural Resources, Wildlife. Technical Report No. 99-02W.

Hilderman Witty Crosby Hanna & Associates, Johnson & Weichel and E.G. Walker. 1996. "East Bank South Development Plan." Submitted to Meewasin Valley Authority, Vol. I and II.

Houston, W.S.L. 2000. "Landscape Classification and Impact of Cattle Grazing on vVegetation and Range Condition in the Dundurn Sand Hills, Saskatchewan." M.Sc. Thesis, Department of Plant Sciences, University of Saskatchewan, Saskatoon, Saskatchewan.

IPCC WGI. 2001. "Summary for Policymakers, Climate Change 2001: The Scientific Basis." Third Assessment Report of Working Group I of the Intergovernmental Panel on Climate Change. (http://www.ipcc.ch/pub/spm22-01.pdf).

——. 2001b. "Summary for Policymakers, Climate Change 2001: Impacts, Adaptation, and Vulnerability." Third Assessment Report of Working Group II of the Intergovernmental Panel on Climate Change. (http://www.ipcc.ch/pub/wg2SPMfinal.pdf).

Lone Pine Resources Ltd. 1990. "Environmental Protection Plan, Freefight Millie Proposed Gas Production Project, Phase 3 and 4, Great Sand Hills, Saskatchewan." (two volumes with addendum).

Macdonald, I.D. 1997. *Vascular Plant Flora Component Report—Canadian Forces Base Suffield National Wildlife Area, Wildlife Inventory.* Edmonton: Canadian Wildlife Service, Environment Canada.

Manitoba Natural Resources Parks. 1995. "Spruce Woods Provincial Park—Backcountry Plan." Unpublished Report.

Manitou Sand Hills Planning and Advisory Committee. 1996. *Manitou Sand Hills Integrated Resource Management Plan.* Regina: Saskatchewan Environment and Resource Management, Saskatchewan Agriculture and Food.

Marr Consulting and Communications Ltd. 1995. "Management Strategies for Sandhill Communities on Crown Land in Southwestern Manitoba." Prepared for Manitoba Natural Resources, Manitoba Habitat Heritage Corporation, Critical Wildlife Habitat Program and the Canadian Wildlife Service.

Muhs, D.R. and P.B. Maat. 1993. "The Potential Response of Eolian Sands to Greenhouse Warming and Precipitation Reduction on the Great Plains of the U.S.A.," *Journal of Arid Environments* 25: 351–361.

Muhs, D.R. and S.A. Wolfe. 1999. "Sand Dunes of the Northern Great Plains of Canada and the United States." Pp. 183–197 in D.S. Lemmen and R.E. Vance (eds.), *Holocene Climate and Environmental Change in the Palliser Triangle: A Geoscientific Context for Evaluating the Impacts of Climate Change on the Southern Canadian Prairies.* Geological Survey of Canada, Bulletin 534.

NOAA 2004. National Climate Data Center archives. http://www.ncdc.noaa.gov/pub/data/normals. (Last Updated February 24, 2004. Accessed November, 2003).

Pope, V.D., M.L. Gallini, P.R. Ronwtree and R.A. Stratton. 2000. "The Impact of New Physical Parameterizations in the Hadley Centre Climate Model—HADCM3," *Climate Dynamics* 16: 123–146.

Prairie Farm Rehabilitation Administration. 1998. *Detailed Ggrazing Report and Survey of Rangeland Condition on PFRA Grazing Land Within the Suffield Military Reserve, Department of National Defense.* Regina: Prairie Farm Rehabilitation Administration.

Ramaley, F. 1939. "Sand-hill Vegetation of Northeastern Colorado." *Ecological Monographs* 9: 1–51.

Reynolds, H.W., S.J. Barry and H.P.L. Kiliaan. 1999. *Small Mammal Component Report—Canadian Forces Base Suffield National Wildlife Area, Wildlife Inventory.* Edmonton: Canadian Wildlife Service, Environment Canada.

Roeckner, E., K. Arpe, L. Bengtsson, M. Christoph, M. Claussen, L. Dümenil, M. Esch, M. Giorgetta, U. Schlese and U. Schulzwelda. 1996. "The Atmospheric General Circulation Model ECHAM-4: Model Description and Simulation of Present-Day Climate." Max-Planck Institute for Meterology Report No. 218, Hamburg Germany.

Sala, O.E., W.J. Parton, L.A. Joyce and W.K. Lauenroth. 1988. "Primary Production of the Central Grassland Region of the United States," *Ecology* 69: 40–45.

Saskatchewan Environment and Public Safety (SEPS). 1991. *Great Sand Hills Land Use Strategy.* Saskatoon: Saskatchewan Environment and Public Safety.

Saskatchewan Environment and Resource Management (SERM). 1999. *Fort à la Corne Integrated Forest Land Use Plan: Background Information.* Saskatoon: Saskatchewan Environment and Resource Management.

——. 2000. "Nisbet Provincial Forest Integrated Forest Land Use Plan." Background Document. Saskatchewan Environment and Resource Management, Forest Ecosystems Branch.

Schimel, D.S., W.J. Parton, T.G.F. Kittel, D.S. Ojima and C.V. Cole. 1990. "Grassland Biogeochemistry: Links to Atmospheric Processes." *Climatic Change* 17: 13–25.

Schykulski, K. and J. Moore. 2000. "Spruce Woods Provincial Park—Prairie Management Plan." Manitoba Natural Resources Parks.

Shandruk, L.J., D.W. Ingstrup, H. Armbruster and S. Barry. 1998. *Ungulate Component Report—Canadian Forces Base Suffield National Wildlife Area, Wildlife Inventory.* Edmonton: Canadian Wildlife Service, Environment Canada.

Sims, P.L., J.S. Singh and W.K. Lauenroth. 1978. "The Structure and Function of Ten Western North American Grasslands: I. Abiotic and Vegetational Characteristics," *Journal of Ecology* 66: 251–285.

Thornthwaite, C.W. and J.R. Mather. 1970. "Instructions and Tables for Computing Potential Evapotranspiration and the Water Balance," *Publications in Climatology* 10: 185–311.

Thorpe, J.P. 1990. "An Assessment of Saskatchewan's System of Forest Site Classification." Saskatchewan Research Council Publication No. E-2530-1-E-90.

Thorpe, J. and R. Godwin. 1993. "Vegetation Survey of the Manito Sand Hills." Saskatchewan Research Council Publication No. E-2550-1-E-93.

——. 1997. "Forage Use by Deer and Cattle in the Great Sand Hills of Saskatchewan." Saskatchewan Research Council Publication No. R-1540-1-E-97.

Thorpe, J., S.A. Wolfe, J. Campbell, J. LeBlanc and R. Molder. 2001. "An Ecoregion Approach for Evaluating Land Use Management and Climate Change Adaptation Strategies on Sand Dune Areas in the Prairie Provinces." Saskatchewan Research Council Publication No. 11368-1E01.

Tuhkanen, S. 1980. "Climatic Parameters and Indices in Plant Geography," *Acta Phytogeographica Suecica*, Almqvist and Wiksell International, Stockholm.

Wallis, C. and C. Wershler. 1988. *Rare Wildlife and Plant Conservation Studies in Sandhill and Sand Plain Habitats of Southern Alberta.* Edmonton: Alberta Forestry, Lands and Wildlife; Alberta Recreation and Parks and World Wildlife Fund Canada.

Weaver, J.E. 1965. *Native Vegetation of Nebraska.* Lincoln: University of Nebraska Press.

Western Oilfield Environmental Services Ltd. 1997. "Environmental Protection Plan: Heavy Oil Drilling Project LSD 01-36-42-26-W3M, Winter, Saskatchewan." Prepared for Wascana Energy Incorporated.

Wolfe, S.A. 1997. "Impact of Increased Aridity on Sand Dune Activity in the Canadian Prairies," *Journal of Arid Environments* 36: 421–432.

Wroe, R.A., S. Smoliak, B.W. Adams, W.D. Williams, and M.L. Anderson. 1988. *Guide to Range Condition and Stocking Rates for Alberta Grasslands.* Edmonton: Alberta Agriculture, Food and Rural Development, Public Land Management.

CHAPTER 10

Drought, Climate Change, and the Risk of Desertification on the Canadian Prairies

Dave Sauchyn, Sam Kennedy and Jennifer Stroich

Introduction

The least precipitation (<330 mm) in the western interior of North America occurs on the northern Great Plains, including a significant area in southeastern Alberta and southwestern Saskatchewan that comprises the Mixed Grassland ecoregion (Figure 1). This semi-arid to sub-humid ecoregion is at risk of desertification or "Land degradation in arid, semi-arid and dry/sub-humid areas, resulting from various factors, including climatic variations and human impact" (UNEP, 1994). Land degradation has been defined as the "reduction or loss of the biological and economic productivity and complexity of terrestrial ecosystems" (Reynolds, 2001: 61). It has been extensively researched for the northern Great

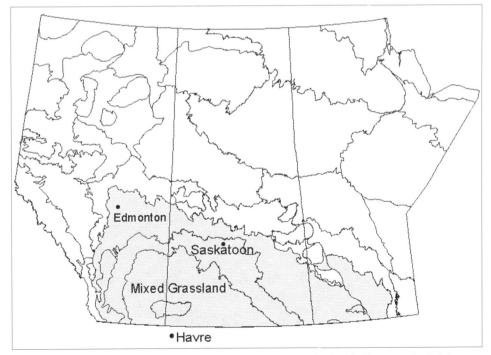

Figure 1. The ecoregions of the Canadian Prairie Provinces. The Prairie Ecozone, shaded in grey, includes the semiarid to subhumid mixed grassland ecoregion.

Plains from the perspective of soil erosion (e.g., Mermut et al., 1983; Pennock and de Jong, 1990; Muhs and Holliday, 1995; Phillips, 1999), range management (e.g., Olson et al., 1985), soil conservation (e.g., Merrill et al., 1996; Peterson et al., 1998; Zhang and Garbrecht, 2002), climate change and drought (e.g., Clark et al., 2002; Soule, 1995), and agricultural policy and sustainable development (e.g., Wilhite and Wood, 1995; Wilson and Tyrchniewicz, 1995). However, the term desertification is rarely, if ever, used to the describe land degradation in this region, perhaps because desertification has historically been mostly associated with famine and the spreading of subtropical deserts (Dregne, 1983; Reynolds and Smith, 2002).

Canada is a party to the United Nations Convention to Combat Desertification (UNCCD). As a donor party, Canada's main responsibility is to support developing countries in their efforts to prevent and mitigate desertification (CIDA, 2000). However, Canada is also considered an affected party because of the risk of desertification over an area of approximately 200,000 km^2 in the prairie provinces. As part of the Mixed Grassland ecoregion (Figure 1), the area at risk represents a small portion (3%) of Canada's total area, but then only a relatively small proportion of the Canadian land mass is arable land. The semi-arid and sub-humid landscapes of the prairie provinces account for over 50% of Canadian farm cash receipts and represent over 80% of the nation's agricultural land. Any loss of productivity and biodiversity in this region has social, economic, and environmental implications for Canada as a whole.

There is a long history of research aimed at enhancing and sustaining the productivity of the agricultural lands of the Canadian Prairies. Concurrent with the Euro-Canadian settlement of the prairies from the 1890s to the early 1900s, a network of experimental farms was established to develop dryland framing practices to prevent wind erosion and mitigate the impacts of drought. The first Canadian government programs to combat land degradation were established in response to the disastrous experience of the 1930s. The federal government created the Prairie Farm Rehabilitation Administration (PFRA) to identify, develop, and implement programs for improving farm practices and land use within the prairie provinces.

Prairie farming practices have historically been adapted to climatic variability to take maximum advantage of soil landscapes:

> Since the settlement of the Prairies in the 19th and early 20th centuries, land use and farming practices have evolved to match the various climates and soil types on the Prairies and adapted to changing markets, technology and transportation systems. The abandonment of farms in the Special Areas of Alberta during the early 1920s, and southwestern Saskatchewan in the 1930s, provides evidence of these adjustment processes. More recently, since the 1980s, there has been a reduction in the summerfallow and an expansion of crop varieties, particularly in areas of higher moisture. (PFRA, 2000: 81)

In this region, the producers and agricultural institutions have demonstrated the capacity to adapt to variation in climate and water resources (Hill and Vaisey, 1995; Riemer, this volume), although prairie agri-ecosystems and soil landscapes remain sensitive to year-to-year climate variability and the affects of sustained drought on surface and shallow subsurface water balances (Lemmen and Vance, 1999). For example the PFRA (2000) reported that

> The reduction of fall tillage and summerfallow and the adoption of direct seeding systems have decreased the period during which soils are exposed to a high erosion risk. However, there remains ample opportunity for erosion to occur. Severe and widespread erosion can take place during extreme weather events (high winds and heavy rains), and particularly during years of consecutive droughts. (PFRA, 2000: vi)

Thus, the sustainability of prairie agriculture depends on the continuous adjustment of land use and management systems to climatic variability, the periodic fluctuation of atmospheric conditions (e.g., drought, early frosts, and storms), and to climatic change, a significant departure from previous average conditions. The agricultural sector is now faced with the prospect of climatic change that may include climatic variability that exceeds the historic experience. This potential shift in climate outside of the previous experience of the prairie agricultural community is coupled with uncertainty in the social and economic future, and evidence that national economic policy does not favour prairie agriculture to the extent that it did in the early- to mid-1900s (Knutilla, 2003).

Here we argue that, although modern soil conserving farming practices can prevent land degradation throughout most of the sub-humid and semi-arid region of the Canadian Prairies, climate change presents a scenario for the 21st century that is much more challenging for the prevention of desertification. The most plausible climate future for the Canadian Prairies, based on the latest global climate models, includes increased potential for evapotranspiration with higher average temperatures. If the greater evaporative water loss exceeds the potential increases in precipitation, then the region can expect declining water supplies associated with droughts of greater severity or longer duration. This paper examines the changing risk of desertification with climate change and more severe drought.

Desertification

Because there is a lack of agreement about the processes that comprise desertification and about the goals of programs for the study and mitigation of land and vegetation degradation in dry environments, there are more than 100 definitions of desertification (Reynolds, 2001). Disagreements and misconceptions can be attributed to the complexity of the desertification process, which involves the interaction of biophysical and socio-economic factors over a range of spatial and temporal scales (Blaikie, 1985; Dregne, 1983; Heathcote, 1983; Reynolds and Smith, 2002; Schlesinger et al., 1990). The biophysical processes of desertification can be measured on a range of scales, from metre-scale plots to atmospheric circulation (Middleton and Thomas, 1992), and social factors vary from grassroots social actions (Hambly and Angura, 1996) to global economic factors, political agreements and international treaties (e.g., the Kyoto accord).

While most definitions of desertification are variations on a central theme of land degradation, usually in dry environments, the meaning ranges from simply the expansion of deserts to the entire suite of processes of land and vegetation degradation, including waterlogging of soils from excessive irrigation. A chronology of definitions reveals the evolution of thinking and changing context. The term desertification was first used to by Aubreville (1949) to refer to desert-like conditions created by erosion processes. This original definition has been paraphrased as "the spreading of deserts" because desertification was first recognized in northern Africa. Drought and famine in the same region during 1968–73, and the global reaction to this crisis, brought desertification into the public and political arena (Glantz, 1987) resulting in the first UN conference on desertification in 1977. In this context, the definition of desertification arising from this conference emphasized the loss of soil productivity:

> The reduction or destruction of the biological potential of land that can lead to desert-like situations. It is an aspect of ecosystem degradation following a consistent reduction in their biological potential... (UNEP, 1977: 6)

Subsequent UNEP definitions (FAO/UNEP, 1984; UNEP, 1991) have included the role of human action and climate as agents of desertification. The current UNEP definition is given in Article 1(a) of the UNCCD (UNEP, 1994: 1334): "Land degradation in arid, semi-arid and dry/sub-humid areas, resulting from various factors, including climatic variations and human impact." This is the definition adopted for the UNCCD by all parties to the convention and thus is the meaning of desertification applied in this paper. This definition excludes the degradation of humid ecosystems and the driest of environments (hyper-arid) because they are natural deserts. The world's arid, semi-arid and dry sub-humid regions have low relative productivity but large total global productivity given their geographic extent (about 40% of the Earth's land area). This low relative productivity renders dryland ecosystems vulnerable to degradation, particularly during periods of drought.

The UNCCD definition refers explicitly to climatic variation. Aridity is the defining characteristic of drylands and varies in degree with climate change. Drought is an aspect of climatic variability: short-term departures from average conditions. Drought can result in some land degradation, and thus tends to raise awareness of desertification, but dryland ecosystems normally have sufficient resilience to recover from drought. In the context of land degradation, resilience is the property of ecosystems that makes them sustainable relative to their use (Warren and Agnew, 1988). A fundamental characteristic of desertification is that human activities accelerate processes of degradation until the soil and vegetation can no longer sustain the current land use without a change in climate or productivity. Severely degraded landscapes may essentially never recover in terms of sustaining the original land use and productivity (Mouat et al., 1997).

Land can be naturally degraded by water and wind erosion, salinization, compaction, and the disturbance of vegetation, but these are not the root causes of desertification. Desertification is land degradation that is accelerated by human activities to such an extent that the current land use is unsustainable, resilience is lost, and the loss of productivity is irreversible. Any human action that contributes to land degradation in drylands is a cause of desertification (Dregne, 1983; Heathcote, 1983; Thomas and Middleton, 1994). These include direct causes like overgrazing, excessive cultivation, poor irrigation practices, deforestation and industrial activities, and coarser-scale factors like overpopulation, governmental policies (agricultural, economic, and environmental), migration, levels of local environmental knowledge, sustainability of communities, and changing economic markets. There are so many direct and root causes that authors have compiled them into categories. Reynolds (2001) organizes them into: overexploitation; destruction of habitat; fragmentation of habitat; introducing and spreading exotic organisms; air, soil, and water pollution; and global warming. Thomas and Middleton (1994) and Warren and Agnew (1988) classify them as population density and change; underdevelopment, poverty, and inequality; commonly held property; attitudes, perceptions and values; internationalism (globalization); colonialism and imperialism; ignorance and inexperience; inappropriate technology and advice; and war and civil unrest.

Collectively and cumulatively, these human activities extract goods and services from ecosystems, altering them to agro-ecosystems. These demands can exceed the carrying capacities of dryland agro-ecosystems resulting in land degradation, especially when climate change and variability limit productivity and resilience. Because rural people are usually the most directly dependent on the local agro-ecosystem by harvesting ecosystem services, they are often most vulnerable to regional changes in the landscape (Blaikie, 1985; Blaikie and Brookfield, 1986). On the Canadian Prairies, for example, rural regions are not necessarily economically poor but they are the first to be negatively influenced by the degradation of natural capital.

From the global inventory of the human dimensions of desertification, a shorter list of social and economic factors is applicable to Canada, including national and provincial agricultural and economic policy; global markets and commodity prices; demographics; and declining rural population. Yet Canadian agricultural institutions and producers have demonstrated the adaptive capacity and social capital to prevent widespread desertification. This has been achieved with various government programs, notably the PFRA following the dust bowl of the 1930s, and progressive adoption of soil- and range-conserving practices. Average Saskatchewan wheat yields from 1909 to 2002 shows wide fluctuations over the first several decades followed by low yields during the 1930s (Figure 2). Since 1940, however, there has been a significant and consistent increase in average yield. These data also indicate that yield has become less variable from year to year, especially since the 1960s, excluding the years of severe drought: 1961, 1988, and 2001–02. This decrease in inter-annual variability is evident when the wheat yields are plotted as departures from expected yields (Figure 3). Expected yield was defined by calculating the mean yield during the period between 1908 and 1939, then as the yield predicted from a least squares regression based on the strong linear trend from 1940 to 2002. Also plotted in Figure 3, as departures from the mean, is annual precipitation for Saskatoon, which is located in the heartland of the Saskatchewan grain belt (Figure 1). The impact of sustained drought on wheat production is evident where large, negative departures from expected yield correspond to two or three consecutive years of low precipitation.

The data in Figures 2 and 3 suggest that prairie farmers have shown a persistent capacity to adapt to variability in climate and economic circumstances by achieving progressively higher and more consistent cereal crop yields, while protecting more land from degradation. A discussion of the input and environmental costs of this adaptation is beyond the scope of this article (but see Stirling, this issue). However, with better soil, water, and crop management, the production of cereal crops has become less vulnerable to climate variability, although not to sustained drought. Within the last 20 years, different cropping systems and the adoption of soil conservation practices, such as reduced tillage and zero-till, have begun to reverse the decline in soil productivity on over one-third of the annually

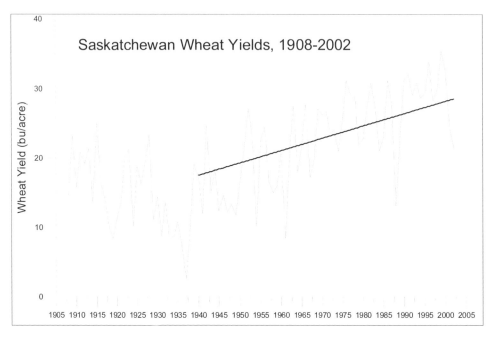

Figure 2. Wheat yields in Saskatchewan from 1908 to 2002. Note the impact of the 1930s and the significant rise in yields since then. Yield also has become more consistent over time, excluding the impact of severe drought in 1961, 1988, and 2001–02. The crop yield data were downloaded from Statistics Canada (http://www.statcan.ca).

cropped land of the prairies. Acton and Gregorich (1995) estimated that the implementation of soil conservation practices resulted in a 7% decrease in the risk of wind erosion and an 11% decrease in the risk of water erosion between 1981 and 1991. More recent statistics (McRae et al., 2000) indicate a 32% reduction in the risk of wind erosion in the prairie provinces between 1981 and 1996 and a 26% increase for the same period in the area of cropland in Saskatchewan at risk of tolerable levels of water erosion.

Thus, land degradation is preventable throughout most of the sub-humid and semi-arid region of the Canadian Prairies given recent climatic conditions, the current policy framework, and crop and soil management regimes. Desertification is still an issue, however, because crops and soil conservation can fail with extreme or prolonged drought, the most prolonged drought in the historical record recently occurred over much of the western Canadian Prairies, and most of the relevant climatic and social variables are moving in directions that increase the risk of land degradation. This paper concludes below with a discussion of the climate variables. With respect to the social variables, Knutilla (2003) documents the declining national significance of prairie agriculture since the 1960s. This was due to the rising influence of global market forces and transnational corporations, that promise economic strategies that span North America with resulting loss of local economic sovereignty, coupled with declining rural population, reduced federal government support for agricultural research, farm crisis relief, income support programs and grain transportation. Furthermore, more powerful and automated farm equipment and larger scales of production enable fewer producers to produce more commodities. This industrial scale of production, and the associated corporate

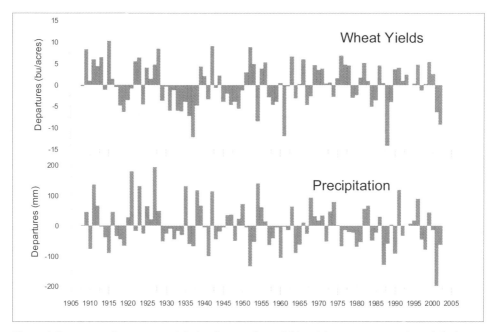

Figure 3. Departures from expected Saskatchewan wheat yield and from mean annual precipitation at Saskatoon (Figure 1) for the period from 1908 to 2002. Wheat yield has become less variable from year to year, with the exception of low yields during the most severe drought, including 2001–02. The impact of sustained drought is also evident in that some of the largest negative departures from expected yield occur during the second or third year of low precipitation. Expected yield was defined as the mean during the period from 1908 to 1939 and then as the yield predicted from least squares regression from 1940 to 2002, given the strong, linear trend in yield as illustrated in Figure 2.

attitudes and values, may present more limited options for prevention of desertification and adaptation to climate change than a more robust, viable, and cohesive network of rural communities (Diaz et al., 2003).

Climate Change, Drought, and Desertification

Experiments using global climate models (GCM) based on elevated greenhouse gases forecast an increase in winter precipitation for the northern Great Plains, decreased net soil moisture and water resources in summer, and more frequent extreme departures from mean conditions, including severe drought (Boer et al., 2000; Hengeveld, 2000; Kharin and Zwiers, 2000). Plate 1 shows a suite of climate change scenarios derived from various global climate models and emission scenarios for the Prairie Ecozone. This plot was generated at the web site (www.cics.uvic.ca/scenarios/index.cgi) of the Canadian Climate Impacts Scenarios project, where the reader can find an explanation of the climate models and emission scenarios. The median increases in mean annual temperature and total annual precipitation are about 3° and 5%, respectively. These changes are significant and the models predict higher evapotranspiration but a marginal increase in precipitation, resulting in a net decrease in available moisture for plant growth.

A scenario of increased aridity and more severe drought has major implications for the risk of land degradation in the southern Canadian Prairies. The link between aridity and erosion is well established from paleo-environmental records (Wolfe et al., 2001, Last and Vance, 2002) and from the monitoring of erosional processes and regional sediment yields (Knox, 1984). Less protection of the soil surface from wind and rain is generally given or implied as the cause of higher rates of erosion in semi-arid landscapes. Plants also reduce runoff erosion through the transpiration of soil water and because stems, roots, and organic matter contribute to the infiltration of rain and snow melt water (Thornes, 1985).

The soil-water balance, the primary control on the sensitivity of soil landscapes to climate, can be modeled at a regional scale as the ratio of precipitation to potential evapotranspiration (P/PET). Semi-arid and sub-humid landscapes, those at risk of desertification, are defined by a P/PET of less then 0.65 (Middleton and Thomas, 1992). Plate 2 shows a map of P/PET using instrumental temperature and precipitation data for 1961–90 and outputs from the Canadian GCM2 (emission scenario B2) for the 2050s (2049–60). The derivation of P/PET and gridded climate data is described in Sauchyn et al. (2002). In Plate 2, these data are mapped by soil texture to show the distribution of the light (sand and sandy loam), and generally most erodable soils, relative to the climate risk. The results illustrate that by the 2050s, there may be approximately 50% more land at risk of desertification.

As with most coupled natural and social systems, prairie landscapes and agri-ecosystems are disturbed not by trends in temperature or precipitation (climate change), but by extreme events and short-term departures from average conditions (climatic variability). Increased aridity on the Canadian Prairies most likely will be realized by more frequent and/or sustained negative departures from a mean water balance. During drought years, semi-arid and sub-humid conditions occur on large parts of the Canadian Prairies, as illustrated in Plate 3 using 2001 data. During 2000–03, most climate stations in Alberta and western Saskatchewan (e.g., Edmonton and Saskatoon, Table 1) had the least precipitation for any four-year period in the instrumental record. This recent drought is possibly more characteristic of prairie climate than the droughts of the 20th century, which were less sustained but still costly in terms of social cohesion in the 1930s, and financial relief in the 1980s (Nemanishen, 1998; Wheaton and Arthur, 1989).

The short Euro-Canadian history on the northern prairies has provided science and agriculture with a relatively limited experience with the range of potential regional climate variability. Paleoclimate records (Sauchyn et al., 2003b) indicate that the 20th century had a relatively favourable climate for human activities, because the droughts were of shorter duration than some from the recent past. The Bears Paw Mountains near Havre, Montana, about 100 km south of the Alberta-Saskatchewan boundary, are forested with trees that provide a proxy climate record for an assessment of drought frequency and duration that extends well beyond the instrumental climate record. A tree-ring chronology from Douglas Fir (*Pseudotsuga menziesii*) enables the reconstruction of annual precipitation, from July of the previous year to June of current year, for the period between 1730 and 2002 (Figure 4). The tree-ring model was calibrated and validated using monthly precipitation from 1901 to 2002 measured at Havre, Montana. The statistics listed in Table 2 verify the predictive skill of the model based on a correlation of 0.74 between July-June precipitation and the residual ring-width index chronology.

When reconstructed July-June precipitation is plotted as departures from the median for the entire record (Figure 5), negative departures (drought years) concentrate during sustained periods of consistently below normal precipitation, such as 1840 to 1870. Multi-year drought is lacking during periods of consistently above

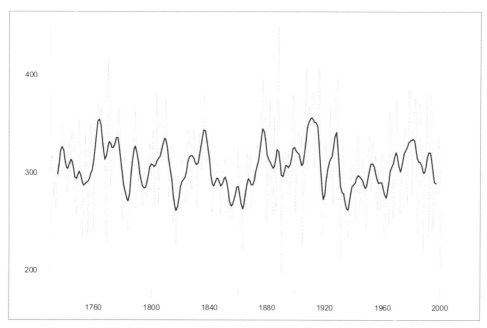

Figure 4. July-June precipitation (mm) for Havre, Montana for the period between 1730 and 2002 as reconstructed from tree-rings from Douglas fir from the Bears Paw Mountains.

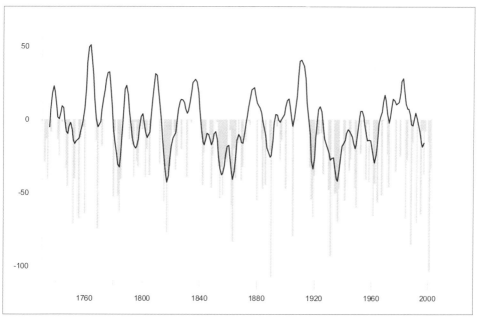

Figure 5. Reconstructed July-June precipitation (mm) shown as departures from the median value for the period of record, 1730 to 2002. This plot illustrates the concentration of drought years during long periods of consistently below normal precipitation, such as 1840 to 1870, and periods of consistently above normal precipitation, when multi-year drought is lacking. These wetter periods include 1890 to 1915, when crop production became established on the western plains, and much of the 1961 to 1990 period, the interval of "normal" climate.

normal precipitation, notably 1890 to 1915 when crop production was established as the dominant economic activity on the western plains, and during 1961 to 1990, which climatologists have used as the interval of "normal" climate. During the many successive dry years of the mid-19th century, John Palliser surveyed the Canadian Prairies and declared a large area "forever comparatively useless" (Palliser, 1859). Whereas drought was frequent in the 20th century and there were some devastating years in that period (1937, 1961, 1988), consecutive years of drought are more common prior to settlement. In most respects, duration is the critical factor, since sustained drought has cumulative impacts and prevents the recovery provided by intervening years of normal to above average precipitation. Not surprisingly, prolonged drought has been implicated as a driving force of landscape change on the northern prairies.

Conclusion

When climate change scenarios are applied to the modeling and mapping of aridity (P/PET) on the Canadian Prairies, the area of land at risk of desertification increases by about 50% between recent conditions (1961–90) and the 2050s. Furthermore, paleoclimate records indicate that during the 20th century this region had a relatively favourable climate for agriculture, because the droughts were generally of short duration. Droughts of longer duration, like those that characterized the pre-settlement history of this region and are forecasted to occur with global warming, are more likely to exceed soil moisture thresholds, leaving agri-ecosystems more vulnerable to disturbance and potentially desertification. The tree-ring reconstruction of precipitation presented here for Havre, Montana reveals that the climate of the northern prairies is more variable than the 20th century records suggest. Even if global warming does not occur as forecasted by GCMs, the paleoclimate data suggest that future climate extremes may exceed those experienced during the 20th century. The most plausible climate future for the northern prairies, based on the latest climate models, includes a declining net surface and soil water balance, as water loss by evapotranspiration potentially exceeds precipitation to a greater degree. This scenario will demand adaptations by society to changes in soil and water systems to limit the risk of desertification.

Despite the vast area and relatively sparse population of the Canadian prairie provinces, most of the landscape is managed for agriculture. The effects of current land use and management practices on rates of surface processes are more immediate than effects due to climate change (Jones, 1993). Thus, agriculture has the potential to significantly mitigate or promote the regional influences of climate. An expansion of agriculture into the currently forested margins of the Prairie Ecozone will require assessment of the sensitivity of these soil landscapes to both climate change and a changed surface cover. Conversely, the soil landscapes of the semi-arid southwestern plains may become more marginal for agriculture and at a greater risk of degradation. While adjustments to farm management practices enable adaptation to climate change, climate change impacts need to be modeled and mapped in a manner that is transferable to local and regional planning and policy making. Conventional approaches to assessment of land degradation, such as simply estimating potential soil loss from fields, will not support the integrated planning over large areas required for adaptation to the impacts of climate change. We must be more proactive.

Acknowledgements

Part of this article was based on a report prepared for the Prairie Farm Rehabilitation Administration (PFRA) on Desertification Indicators. We also thank Antoine Beriault for assistance with the tree-ring research, and Justin Derner (USDA-ARS) and an anonymous reviewer for their helpful comments on the manuscript. Research funding was provided by NSERC, SSHRC and the Climate Change Action Fund.

References

Acton, D.F. and L.J. Gregorich (eds.). 1995. "The Health of Our Soils: Toward Sustainable Agriculture in Canada." Centre for Land and Biological Resources Research Branch, Agriculture and Agri-Food Canada. Publication 1906/EBie, Stein W. 1990. Dryland Degradation Measurement Techniques, World Bank, Environment Work Paper No. 26.

Aubreville, A. 1949. *Climats, forets et Desertification de l'Afrique tropicale*. Paris: Soc. d'éditions géographiques maritimes et coloniales.

Blaikie, P.M. 1985. *The Political Economy of Soil Erosion*. London: Longman.

Blaikie, P.M. and H. Brookfield. 1986. *Land Degradation and Society*. London: Methuen.

Boer, G.J., G. Flato and D. Ramsden. 2000. "A Transient Climate Change Simulation with Greenhouse Gas and Aerosol Forcing: Projected Climate for the 21st Century," *Climate Dynamics* 16: 427–450.

CIDA. 2000. *Desertification—A Canadian Perspective*. Ottawa: Canadian International Development Agency.

Clark J.S., E.C. Grimm, J.J. Donovan, S.C. Fritz, D.R. Engstrom and J.E. Almendinger. 2002. "Drought Cycles and Landscape Responses to Past Aridity on Prairies of the Northern Great Plains, USA," *Ecology* 83, no. 3: 595–601.

Diaz. H.P., J. Jaffe and R. Stirling. 2003. *Farm Communities at the Crossroads: Challenge and Resistance*. Regina: Canadian Plains Research Center.

Dregne, H. 1983 *Desertification of Arid Lands. Advances in Arid Land Technology and Development*. Chur: Harwood Academic Publishers.

FAO/UNEP. 1984. *Provisional Methodology for Assessment and Mapping of Desertification*. Rome: FAO.

Glantz, M.H. 1987 "Drought in Africa," *Scientific American* 256: 34–40.

Hambly, Helen and Tobias Onweng Angura. 1996. *Grassroots Indicators for Desertification*. Ottawa: International Development Research Centre.

Heathcote, R.L. 1983. *The Arid Lands: Their Use and Abuse*. London: Longman.

Hengeveld, H.G. 2000. *Climate Change Digest: Projections for Canada's Climate Future*. CCD 00-01 Special edition. Ottawa: Environment Canada.

Hill, H. and J. Vaisey. 1995. "Policies for Sustainable Development." Pp. 51–62 in D.A. Wilhite and D.A. Wood (eds.), *Planning for a Sustainable Future: The Case of the North American Great Plains. Proceedings of the Symposium, May 8–10, 1995*. Lincoln, NE: n.p.

Jones, D.K.C. 1993. "Global Warming and Geomorphology," *The Geographical Journal* 159, no. 2: 124–130.

Kharin, V.V. and F.W. Zwiers. 2000. "Changes in the Extremes in an Ensemble of Transient Climate Simulations with a Coupled Atmosphere-Ocean GCM," *Journal of Climate* 13: 3760–3788.

Knox, J.C. 1984. "Fluvial Responses to Small Scale Climate Changes." Pp. 318–342 in J.E. Costa and P.J. Fleisher (eds.), *Developments and Applications of Geomorphology*. Berlin: Springer-Verlag.

Knutilla, M. 2003. "Globalization, Economic Development and Canadian Agricultural Policy." Pp. 289–302 in H.P. Diaz, J. Jaffe and R. Stirling (eds.). *Farm Communities at the Crossroads: Challenge and Resistance*. Regina: Canadian Plains Research Center..

Last, W.M. and R.E. Vance. 2002. "The Holocene History of Oro Lake, One of Western Canada's Longest Continuous Lacustrine Records," *Sedimentary Geology* 148: 161–184.

Legates, D.R. and C.J. Willmott. 1990. "Mean Seasonal and Spatial Variability in Gauge-Corrected, Global Precipitation," *International Journal of Climatology* 10: 111–127.

Lemmen, D.S. and R.E. Vance. 1999. "An Overview of the Palliser Triangle Global Change Project." Pp. 7–22 in D.S. Lemmen and R.E. Vance (eds.), *Holocene Climate and Environmental Change in the Palliser Triangle: A Geoscientific Context for Evaluating the Impacts of Climate Change on the Southern Canadian Prairies*. Ottawa: Geological Survey of Canada.

McRae, T., C.A.S. Smith and L.J. Gregorich (eds.). 2000. *Environmental Sustainability of Canadian Agriculture: Report of the Agri-Environmental Indicator Project. A Summary*. Ottawa: Agriculture and Agri-Food Canada.

Mermut, A.R., D.F. Acton and W.D. Eilers. 1983. "Estimation of Soil Erosion and Deposition by a

Landscape Analysis Technique on Clay Soils in Southwestern Saskatchewan," *Canadian Journal of Soil Science* 63, no. 4: 727–739.

Merrill, S.D., A.L. Black and Bauer, 1996. "Conservation Tillage Affects Root Growth of Dryland Spring Wheat under Drought," *Soil Science Society of America Journal* 60, no. 2: 575–583.

Middleton, N. and D.S.G. Thomas. 1992. *World Atlas of Desertification.* London: United Nations Environment Program.

Mouat, D.L., J. Lancaster, T. Wade, J. Wickham, C. Fox, W. Kepner and T. Ball. 1997. "Desertification Evaluated using an Integrated Environmental Assessment Model," *Environmental Monitoring and Assessment* 48: 139–156.

Muhs, D.R. and V.T. Holliday. 1995. "Evidence of Active Dune Sand on the Great Plains in the 19th Century from Accounts of Early Explorers," *Quaternary Research* 43, no. 2: 198–208.

Nemanishen. W. 1998. *Drought in the Palliser Triangle* (a provisional primer). Regina: Prairie Farm Rehabilitation Administration.

Olson, K.C., R.S. White and B.W. Sindela. 1985. "Response of Vegetation of the Northern Great Plains to Precipitation Amount and Grazing Intensity," *Journal of Range Management* 38, no. 4: 357–361.

Palliser, J. 1859. *Papers Relative to the Exploration of That Portion of British North America Which Lies Between the Northern Branch of the River Saskatchewan and the Frontier of the United States; and Between the Red River and the Rocky Mountains.* New York: Greenwood Press.

Pennock, D.J. and E. de Jong. 1990. "Spatial Patterns of Soil Redistribution in Boroll Landscapes, Southern Saskatchewan, Canada," *Soil Science* 150, no. 6: 867–873.

Peterson G.A., A.D. Halvorson, J.L. Havlin, O.R. Jones, D.J. Lyon and D.L. Tanaka. 1998. "Reduced Tillage and Increasing Cropping Intensity in the Great Plains Conserves Soil C," *Soil and Tillage Research* 47, nos. 3-4: 207–218.

PFRA. 2000. *Prairie Agricultural Landscapes: A Land Resource Review.* Regina: Prairie Farm Rehabilitation Administration.

Phillips, S.T. 1999. "Lessons from the Dust Bowl: Dryland Agriculture and Soil Erosion in the United States and South Africa, 1900–1950," *Environmental History* 4, no. 2: 245–266.

Reynolds, J.F. 2001. "Desertification." Pp. 61–78 in *Encyclopedia of Biodiversity*, Volume 2. Academic Press.

Reynolds, J.F. and M.S. Smith (eds.). 2002. "Do Humans Cause Deserts?" In *Global Desertification: Do Humans Cause Deserts?* Berlin: Dahlem University Press.

Sauchyn, D.J., E. Barrow, R.F. Hopkinson and P. Leavitt. 2004. "Aridity on the Canadian Plains," *Géographie physique et Quaternaire* 56, nos. 2-3: 247–259.

Sauchyn, D.J., J. Stroich and A. Beriault. 2003b. "A Paleoclimatic Context for the Drought of 1999–2001 in the Northern Great Plains," *The Geographical Journal* 169, no. 2: 158–167.

Schlesinger, W.H., J.F. Reynolds, G.L. Cunningham, L.F. Huenneke, W.M. Jarrell, R.A. Virginia and W.G. Whitford. 1990. "Biological Feedbacks in Global Desertification," *Science* 247: 1043–1048.

Soule, P.T. 1995. "Drought Evolution Patterns in the USA during Great Plains-Centered Droughts," *Great Plains Research* 5, no. 1: 115–135.

Thomas D.S.G. and N.J. Middleton. 1994. *Desertification; Exploding the Myth.* Chichester: John Wiley & Sons.

UNEP. 1977. *Desertification, Its Causes and Consequences.* Oxford: Pergamon Press.

——. 1991. *Status of Desertification and Implementation of the United Nations Plan of Action to Combat Desertification.* Nairobi: n.p.

——. 1994. *United Nations Convention to Combat Desertification in those Countries Experiencing Drought and/or Desertification, Particularly in Africa.* Geneva: n.p.

Warren, A. and C. Agnew. 1988. *An Assessment of Desertification and Land Degradation in Arid and Semi-arid Areas.* London: Ecology and Conservation Unit, University College.

Wheaton, E.E. and L.M. Arthur. 1989. *Some Environmental and Economic Impacts of the 1988 Drought.* Sasktoon: Saskatchewan Research Council.

Wilhite, D.A. and D.A. Wood (eds). 1995. *Planning for a Sustainable Future: The Case of the North American Great Plains.* IDIC Technical Reprt Series 95-1.

Wilson, Art and Allan Tyrchniewicz. 1995. *Agriculture and Sustainable Development: Policy Analysis on the Great Plains.* Winnipeg : International Institute for Sustainable Development.

Wolfe, S.A., D.J. Huntley, P.P. David, J. Ollerhead, D.J. Sauchyn and G.M. Macdonald. 2001. "Late 18th Century Drought-induced Sand Dune Activity, Great Sand Hills, Southwestern Saskatchewan," *Canadian Journal of Earth Sciences* 38, no. 1: 105–117.

Zhang X.C.J. and J.D. Garbrecht. 2002. "Precipitation Retention and Soil Erosion under Varying Climate, Land Use, and Tillage and Cropping Systems," *Journal of the American Water Resources Association* 38, no. 5: 1241–1253.

CHAPTER 11

Managing Changing Landscapes on the Northern Prairies: Using Functional Groups and Guilds

Todd Radenbaugh

Introduction

The northern prairie landscape was structured primarily by stochastic perturbations in climate (e.g., drought, variations in rainfall and temperature, and unseasonable frost), as well as by fire frequency and grazing of large, keystone herbivores (primarily bison) prior to the introduction of the railways in the 1880s. However, with expanding agricultural and urban settlements, human activities have become an important structuring force at both local and regional levels by adding and managing species and assemblages (Samson and Knopf, 1996; Radenbaugh, 1998). These new human influences have minimized the ecological roles played by climate, grazing, and fire. The activities have converted prairie to cropland and subdivided the landscape into mosaics of different plant assemblages (Plate 1). Now, most of this system is annually maintained by the intentional removal of specific plants (e.g., weeds) and the introduction of exotics (e.g., crops), large-scale annual inputs of chemicals (fertilizers and pesticides), and the active control and management of livestock. This has resulted in fundamental change in how the regional ecosystems operate in terms of nutrient cycling (Matson et al., 1997).

Agriculture has become the dominant land use in the northern prairies. The management of this landscape has generally been guided by the principle of maximizing rangeland and grain production. Thus, the region has become a major food-producing area for North America, making up 60% of total cropland and 80% of the pasture and rangeland in Canada. On the prairies, agriculture, and its processing and other associated industries, is a multi-billion dollar business that has transformed the region into an agro-ecosystem. Thus, the northern prairies are no longer a natural system but an ecosystem extensively influenced by society with large segments dependent on agricultural production. So we must ask: have species and biotic assemblage-level changes transformed the functional components of the broader northern prairie ecosystem? If so, are we harming the ecological services that this ecosystem provides to society?

Society's Influences on Prairie Ecosystems

The influences of society on the northern prairie ecosystems have been investigated extensively. However, most studies have focused primarily on individual endangered or threatened species, such as bison (*Bison bison*) or burrowing owl (*Athene cunicularia*), and unique or important plant assemblages (wildlife habitat such as badlands, remnant grasslands, and wetlands). This is perhaps because the most visible signs of agricultural impacts have been the loss of prominent species and habitats. Nonetheless, the wholesale replacement of native plant cover with crops and pastures is influencing the entire ecosystem, not just individual species and populations or specialized habitats. To examine the changing functions of entire ecosystems, there is a need for investigations into a hierarchy of units that identify the roles of not only species and biotic assemblages but the entire unit (ecosystem) as well.

Rapport et al. (1985) look at a variety of systems and identify many symptoms of stress in an ecosystem, most of which are related to maintaining species assemblages in an earlier successional sere. The signs of stress include decreases in nutrient pools, losses in primary productivity, and lowering of species diversity, as well as the reversion of succession in plant assemblages. In the northern prairies, all of these changes have occurred. One exception is species diversity. There has been a decrease in diversity at the local species assemblage level for grassland birds, but a significant increase has occurred at the broader ecosystem level (Radenbaugh, 2003).

In examining food web structure in ecosystems, Persson et al. (1996) identify four factors that, if altered, modify the dynamics of ecological systems: habitat structure, disturbance fluctuations, population size and structure, and animal behavior and defenses. All of these ecosystem-level influences can be seen in the northern prairies; all are caused, directly or indirectly, by human actions; and all can be linked to agricultural development. Over 70% of the native grassland species assemblages have been significantly altered through tilling and drainage (Samson and Knopf, 1996). Add to this the introduction of shelterbelts, roadways, farmyards, the expansion of woodlands habitats, and increased pesticide use. Furthermore, two major natural disturbance fluctuations (fire and grazing) that maintain native habitat integrity have been changed through the introduction of annual tillage, pest control, fencing, game management, and irrigation. With all of these societal influences, it becomes clear why the structure of nearly all habitats has been affected. But what about the functions at the ecosystem level?

Vegetation Changes

The native vegetation cover in the northern prairies has been altered dramatically since the late 1800s, with agricultural landscapes now dominating the region. Figure 1 shows three snapshots from different databases illustrating the extent to which major native vegetation types have changed in southwest Saskatchewan. In the late 1880s, grasslands covered over 90% of the region, but by the mid-1990s grasslands comprised less than 28% of the area. By the 1920s, cultivated land grew to be the dominant cover, comprising nearly 60% of the region (Canada, 1996; Samson and Knopf, 1996). Furthermore, between the 1930s and 1990s wetland area decreased by approximately 35%, primarily due to the cultivation of marginal lands (Dahl, 1990; Canada, 1996; Samson and Knopf, 1996). In contrast, woodland and brushland areas in this region have increased slightly (by 5%) since the late 1800s due to planting of shelterbelts, farmyards, urban areas, and fire suppression, as is represented in Plate 2 (Canada, 1996; Samson and Knopf, 1996). Another important change is the increase in linear developments, such as roads, power lines, and pipelines, that cross the region. Thorpe and Godwin (1999)

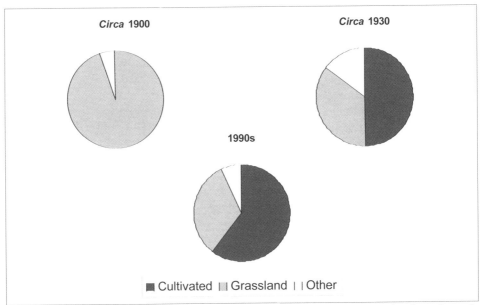

Figure 1. The percentage of landscape in cultivated, grassland, and other vegetation from the early 1880s, 1930s and 1990s in the Mixed Grassland ecoregion of Saskatchewan. The 1880s are estimates from a 1980 database; the 1930s data are from the 1936 Canadian Agricultural Census; and the 1990s data are from the 1996 South Digital Land Survey (PFRA-GIS division).

estimate these features to cover another 2% of the southern Saskatchewan landscape. Far from native prairie, such areas represent native habitat loss, but in the heavily cultivated regions they may be the only important tracks of persistent grassland cover that provide some form of habitat diversity.

Avifauna

The highly mobile nature of bird species allows them to track and select their preferred habitats year to year, allowing the native avifauna to remain in a diverse prairie landscape. Therefore, this fauna makes a good indicator of change in landscapes. Historically, most of the bird species that used the resources in the northern prairies were not specific to the Great Plains, but lived in other grassland regions in North America (Mengel, 1970; Wiens, 1973, 1974, 1989; Cody, 1985). Furthermore, breeding bird population numbers were structured primarily by the limiting amount of resources (mostly the lack of large vertical structures) in homogenous northern grassland and wetland habitats. These habitats also experienced recurrent, but unpredictable climatic "catastrophes" such as drought and unseasonable frosts (Wiens, 1974, 1989) that periodically devastate populations. Successful exploitation of these two-dimensional and climatically severe habitats by breeding birds requires a complex combination of distinct adaptations such as ground nesting, well-defined diversionary displays, flight songs, cryptic coloration, relatively short incubation and nesting periods, and physiological abilities to withstand drought (Wiens, 1974).

Since agricultural settlement, many of the native birds in the northern prairies that have specific prairie adaptations have declined while non-native birds have increased. For example, in areas where agricultural lands dominate, grassland specialist birds show significant negative trends as recorded in the North American

Breeding Bird Surveys (BBS) (Sauer et al., 2000), and relative abundance (Radenbaugh, 2003), especially for species with particular habitat, foraging, nesting, or diet requirements. Furthermore, the habitats that have expanded into the region brought with them different adaptation requirements for birds, in effect changing how species interact with the region.

In the northern prairies, there has been an increase in bird diversity. The majority of this increase is due to species using woodlands, brushlands, and urban habitats (Sauer et al., 2000; Smith and Radenbaugh, 2000; Herkert, 1994; Radenbaugh, 2003). In the mixed grassland of Saskatchewan, the 5% increase in woodland areas allowed species diversity to increase by about 20% (Radenbaugh, 2003). This occurred while agricultural practices have directly reduced grassland habitat by over 60% (Samson and Knopf, 1996; Selby and Santry, 1996). Inevitably, this increase in woodland habitat and birds comes at the expense of the native grassland species. In addition, a comparison of the avifauna between the late 1890s and 1990s shows that mixed grassland assemblages shared only 80% of the same species and that diversity was only 77% similar (Radenbaugh, 2003). Thus, a regional change in the vegetation structure has indirectly altered regional species diversity and potentially the functional components of the prairie ecosystem. As these new habitats have expanded, the broader ecosystem levels have allowed new biota to successfully exploit this landscape. This illustrates that even small changes in vegetation can have profound changes in the regional fauna.

Changing Ecological Functions

When species are added or removed from assemblages, the functioning of the entire ecosystem is often affected (Johnson et al., 1996). We have seen significant changes in species composition in prairie assemblages as a result of agriculture. For example, the avifauna of the 1990s differs markedly from that of the early 1900s (Igl and Johnson, 1997; Houston and Schmutz, 1999; Smith and Radenbaugh, 2000; Radenbaugh, 2003). Since birds are a major component of the ecosystem, changes seen in their resource use (guilds) may mirror changes that have occurred in other populations (mammal predators and competitors, insect prey, and grazed plants). Other major influences in the landscape include:

- reduced local surface water quality—induced by pesticide application on croplands and seasonal sediment loading from poor land management practices (Harker et al., 1997);

- nitrate contamination of groundwater—high phosphate levels in surface waters due to fertilizer applications (Harker et al., 1997; Dixit et al., 2000);

- loss of soil productivity from erosion of topsoil—changes in salinity and pH from irrigation practices and overgrazing (Franzen, 2003; Acton and Gregorich, 1995);

- endangered and threatened species—although the region has few endemic species, there are species at risk. In Saskatchewan 17 animal and 13 plant species have been listed as threatened or endangered (Saskatchewan Conservation Data Centre, 2004) while 38 animals and 5 plants are listed for North Dakota (USFWS, 1995);

- introduction and invasion of exotic plant and animal species—these new species compete with native species for space and resources. They often change the structure and dynamics of local species assemblages and the quality of native habitat for wildlife (Peltzer, 2000);

• increased regional desertification risk—removal of native vegetation coupled with potential increases in temperature and evaporation due to climate change (Sauchyn et al. and Wolfe and Thorpe, this issue).

• Fragmentation of landscapes—less than 35 % of the original vegetation cover remains in scattered islands, with major losses in wetlands. At the local species assemblage level, many of the remaining patches are not large enough to support native area-sensitive species. At the ecosystem level, there are more types of habitats in the 1990s when compared to the native condition. These new habitats contain many invasive plants and animals.

Regional Versus Local Species Diversity

When a plot of land is converted from native prairie to croplands, diversity generally decreases in the field and, to a lesser extent, in the adjacent native fragments that remain. This is because the resource requirements needed by prairie specialists species are no longer present in the altered habitat (Leach and Givnish, 1996; Lesica and DeLuca, 1996; Sutter and Brigham, 1998; Bakker and Berendse, 1999; Christian and Wilson, 1999; Davis and Duncan, 1999; Pepper, 1999). Thus, by altering portions of the landscape, farming promotes species that are better at exploiting agricultural resources, and the remaining plant assemblages can no longer support as many grassland and wetland specialists. This suggests that, at lower species association levels, much of the landscape is becoming more homogeneous ecologically as more land area is cultivated, and there is a direct relationship between local agricultural conversion and lower local species diversity. However, at higher ecological levels, this relationship between habitat loss and diversity is not so clear. The fragmentation of landscapes into smaller patch sizes does not necessarily constrain the regional diversity of plant (Simberloff and Gotelli, 1984) or breeding bird (Friesen et al., 1999) assemblages if there are still large areas of relatively undisturbed habitat blocks in the region. At this broad level, an agricultural landscape will add habitat heterogeneity, thus adding species diversity as well.

Because of scale issues, land-use managers and policy makers need to be cognizant of the many ways species diversity can be used to indicate a system's health. Often, species diversity is used as an indicator of system integrity on the assumption that species diversity enhances productivity and stability (Johnson et al., 1996). However, a higher diversity at the broader scale cannot always be correlated with ecosystem health and habitat quality. For example, on the northern prairies, although the grassland avifauna diversity significantly decreased from 1900 to the 1990s, there has been a significant increase in woodland birds, thus increasing regional diversity (Radenbaugh, 2003). Furthermore, the northern prairie avifauna before agricultural settlement is clearly not the same as today. On the Mixed Grassland ecoregion, the bird assemblages in the early 1900s shared only 108 of the total 135 species with those in the late 1990s and for species diversity they are only 77% similar, according to the Horn Index on Shannon diversity (Radenbaugh, 2003). Another indication of the changes that took place between these periods is the decrease of species ranked as common and rare and an increase in the fairly common and uncommon categories (Figure 2). The historical avifauna was characterized by low diversity with a few species having numerical dominance (Risser et

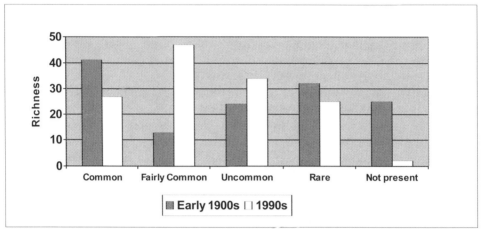

Figure 2. Species richness by relative abundance ranks for all breeding birds in the Mixed Grassland ecoregion of Saskatchewan for two periods: the early 1900s and the 1990s. Early 1900 values are based on abundance ranks assigned to species based on historical accounts (Radenbaugh, 2003).

al., 1981; Cody, 1985; Wiens, 1989), but this has now changed. However, in the mixed grassland of Saskatchewan, the number of these common species has declined while the number of rare species has increased, and this effect is an extension of agricultural influences. Furthermore, 23 species have become naturalized in the region, most of which use habitats that were rare or not available prior to 1900. Thus, if diversity is to be used as an indicator, resource use must also be considered.

Belcher (1958) was one of the first to document this increase in woodland birds in the 1950s near Dilke, SK, observing that the following migrant birds, which were rare in the surrounding grasslands, were commonly nesting in the shelterbelt trees: Western Kingbird, Least Flycatcher, Brown Thrasher, American Robin, Loggerhead Shrike, Warbling Vireo, Chipping Sparrow, Common Grackle, Baltimore Oriole, American Goldfinch, and House Sparrow.

These bird trends indicate that society may have changed the rules that structure the northern prairie species assemblages. The decline in robust grassland generalists, such as the Killdeer and Western Meadowlark, coupled with increases in non-native woodland specialists like Merlin, Veery, Cedar Waxwing, and Chipping Sparrow, suggests that the broad biotic structure of the landscape has changed to the point where one may question if the landscape could still be classified as prairie. Moreover, observations of increased regional habitat diversity at the expense of grassland diversity suggest that there could be cascading effects to other biotic groups and systems, including those upon which society depends. Therefore, conservation plans and policies should address not only the issues of protection of sizeable regions but also the type and use of the landscape. Policy makers and stakeholders in the region must decide what type of regional landscape is important to protect since management practices can have broad influences. Agroecosystem development brings with it different vegetation types that promote or inhibit particular animal species. There is a need to know which species benefit and which lose on a regional scale if we are to manage a healthy ecosystem.

Agricultural Effects

This volume has shown that, despite major conversion of prairie vegetation assemblages to an agricultural system consisting of a variety of land use patterns, many plants and animals will continue to use the resources of the landscape, and an intact ecosystem will persist. Questions remain, however, concerning the health of the system and its ability to provide all the necessary ecological and economic goods and services.

The added landscape heterogeneity and species diversity brought by an industrialized agricultural society is of concern because it alters species composition and ecosystem functioning. These, in turn, alter the ecosystem services provided to society.

In altering the landscape, modern human society has funneled new resources and ecological roles into an ecosystem. This means that local management strategies, planned or not, influence much more than the plot of land or field they are meant to affect. This altering of resources in a landscape has major repercussions for all the evolutionary processes in the region. The greatest immediate repercussion is that modern agriculture has altered food-web interactions and abiotic-resource distribution. Thus, human activities have modified biotic structure even at broad ecological levels, changing the ecological structure and function of the ecosystem. These human influences are diversifying natural selection forces by changing the traits that are selected in the environment and changing the evolutionary trajectory of species. Because of the temporal pace of these ecosystem-level processes, it will require hundreds of years to measure and realize the full extent of evolutionally responses by the biota.

Management Implications

The introduction of agriculture most likely has brought with it a more stable environment. Some of the "catastrophes" such as drought, wildfire, and intense stochastic grazing have been suppressed due to the introduction of irrigation, fire control, and managed grazing. Thus, human influences have significantly altered functional components (e.g., breeding bird guilds) of the ecosystem and are changing the ways in which resources are partitioned for other species.

The prairie landscape has been shown to be anything but in equilibrium with ecological balance being the exception rather then the rule. Natural disturbances need to occur at many levels to add the species assemblage heterogeneity required to maintain landscape-level diversity and ecosystem integrity. There should also be a distinction between agricultural and native land-use strategies. Agricultural areas should be managed to provide maximum short-term economic benefits for landowners while holding changes to the regional ecosystem's integrity to a minimum. Native areas need to be managed to maximize ecosystem functions to maintain the necessary life-support systems and ecosystem integrity as Costanza et al. discuss (1997). The difficult part is to decide how much change is needed to agricultural and native areas to maximize economic, social, and ecosystem health.

Furthermore, habitat loss is not the only important indication of the integrity of the landscape. The way we use a landscape is an important driving force in structuring local species assemblages. As discussed, the use of prescribed fire or managed grazing will alter the grassland species that live there. This is further complicated by the timing, frequency, block size, and intensity for both burns and grazing, since they also have been shown to be important to the survivability of birds (Wright and Bailey, 1982; Rohrbaugh et al., 1999). This can also apply to agricultural practices, including the patchy use of pesticides, crop rotations, tillage practices, timing of cultivation and harvest, and location of shelterbelts. Therefore, the specific species composition in a region depends not only on the overall area of land that is altered but also on land-use policies.

If a decision is made to preserve native grassland habitats, then efforts must be made to ensure that structurally suitable habitat exists in large enough blocks to sustain specialist species. This would mean the elimination of most treed habitats that are important to many non-native species and produce a decline in woodland bird species like kingbirds, doves, bluebirds, sparrows, and flycatchers. Conservation of grasslands should be based on indicator animals with restrictive habitat requirements that require large territories, such as the burrowing owl, Sprague's pipit, prairie falcon, McCown's longspur, chestnut-collared longspur, and yellow-headed blackbird. However, by choosing to promote one suite of habitats, we limit other habitats and structures that promoted the rise in regional bird diversity.

If we are to promote species and abundances that are more similar to the native state, an active approach must be taken to manage whole ecosystems, rather than focusing on policies that are geared for species or species assemblages. Agricultural systems introduce factors that will always influence target species, and this must be acknowledged before society can manage for both agriculture and native grassland habitats. If we continue current practices, the trend towards an increasing woodland avifauna will persist and the non-agricultural regions will evolve into a landscape more similar to a savannah then to a prairie. The regional avifauna will continue to adapt to these changes, with woodland animals increasing at the expense of grassland species. In addition, society must adapt to the changing environment. This is one of the consequences of modern agricultural settlement: humans have come to play an important ecological role in the prairie ecosystem. As one changes, so must the other.

The following are general strategies to manage prairie ecosystems and increase ecosystem services on farms:

- Increase local grassland habitat heterogeneity on farms by varying grass cropping, grazing, burning and mowing regimes.
- Reduce the amount of exotic plant species (e.g., crested wheat grass and smooth brome).
- Promote the use of native prairie species for hay crops and in pastures for extended periods in fields to vary the vegetation cover.
- In areas that are not extensively cropped and where wind erosion is not an issue, limit the expansion of large brushland, woodland, and shelterbelt areas, even though they provide important habitat to many animal species.
- Vary the size and timings of grazing and burning within existing grassland plots. This will vary grassland structure and leads to regional grassland heterogeneity.
- Promote no-till practices that include cultivation early in the bird-breeding season.
- In dryer regions, limit summerfallow practices and, where used, postpone tilling until after mid-summer to allow wildlife to forage and breed.

• Protect wetlands from pesticide and soil runoff by limiting use around wetlands and preventing residue runoff.

• Increase energy efficiency and reduce farm inputs. The easiest way to do this is to reduce non-renewable energy sources as Stirling describes (this issue).

• Promote habitat stewardship that conserves remnant native prairie on private land via conservation easements. Programs such as the Conservation Reserve Program in the US and the Missouri Coteau Initiative in Canada are included in this strategy (Riemer, this issue).

• Involve local groups in the conservation policies of their region as Sutter et al. discuss (this issue).

• Increase the use of conservation consumer-producer cooperatives as Belcher and Schmutz discuss (this issue).

• Increase energy efficiency on farms and introduce low-input farming (Stirling, this issue).

• Manage crop rotations, stocking rates, and timber harvest with an increased attention on potential climate change to reduce negative biological and physical impacts (Wolf and Thorpe, this issue).

• Develop farming techniques and strategies that allow society to adapt to changes in the landscape (Sauchyn et al., this issue) as it is often expensive or impossible to alter the broad scale processes that are the root cause of change.

In native prairie grasslands:

• If native species are the goal, then limit the woodland encroachment within grassland habitats and discourage the construction of artificial habitats, such as nesting boxes outside of farmyards, villages, and cities.

• Allow, or sometimes promote, natural disturbances to operate within native blocks.

• Protect large tracks of native grasslands and wetlands from further agricultural conversion.

• Reduce the amount of invasive and exotic plant species (e.g., crested wheat grass and smooth brome) that are spreading into the remaining native grasslands, since these species change the local species diversity and quality of the native wildlife habitat.

• Increase the size of habitat blocks as grassland specialists generally avoid the edges of habitats.

• Restore converted habitat to grassland cover. Even small blocks (<10 ha) can increase local heterogeneity.

In water and wetlands:

• Protect surface waters from runoff that causes eutrophication (reducing contamination by pesticides and fertilizers) and sedimentation (wind and water erosion).

• Plan and monitor at the watershed level: this will benefit individual users and make decision-making clearer.

• Create buffer zones along rivers to enhance riparian habitat. Such zones will increase water quality by reducing agricultural nutrients, pesticides, and pathogens. Such riparian zones generally provide fish and wildlife with habitats that can benefit pasture management and cattle production (Sutter et al., this volume).

• Recognize surface waters and the quantity and quality of ground water sources, and manage so that both are enhanced.

• Manage wetlands, sloughs, and coulees as a connected water system. All are needed to maintain high water level in the larger lakes and wetlands.

Conservation practices are also more effective when management agencies become integrated and have an overall comprehensive strategy that focuses on all types of capital. This is most effectively done when local groups and stakeholders are involved in partnerships with local and federal agencies.

Conclusions

The influence/footprint of society on the northern prairies is greater now than at any time in recorded history, and it will almost certainly increase. Thus, the northern prairies needs be managed to increase and maximize ecological goods and services for societal and wildlife benefits. Ecological benefits provided to society directly by the prairie landscape include all the services that Costanza et al. (1997) outline—for this list also see Appendix 1 in Radenbaugh and Sutter this issue—including food production and security, wildlife habitat for aesthetic and hunting purposes, soil stability and quality, and clean surface and groundwater. Given the potential effects of climate change, one could add sequestration of atmospheric carbon (Boehm et al., 2000).

Thus, there is a need to measure and monitor ecosystems' functions and diversity. Measures of functional diversity (e.g., guilds) coupled with biodiversity indices would illustrate ecosystem-level integrity. To fully understand how biotic systems function as agro-ecosystems, an integrated approach must be taken that includes input from, and interactions with, society. This means that the lower-level mechanistic studies of species interactions used in the past must be coupled with investigations that examine the broader spatio-temporal levels. Furthermore, more studies that investigate society's role in contributing to regional biodiversity and species distribution are needed. In doing this, we will enhance our understanding of how modern ecosystem functions are structured, allowing improved and more informed management practices.

Since humans are an important functional component of the agro-ecosystem, all sectors of society should be represented and involved in management decisions. Diverse groups and stakeholders need to discuss issues in order to reduce tension between those wanting only to preserve a native ecosystem to maintain ecosystem functions, and those wishing to exploit this region to its fullest economical potential. By exercising controls and zoning, managers may find a middle road satisfactory for users and stakeholders.

In the end, the long-term sustainability of prairie landscapes is reliant on those that live on, and profit from, the landscape. Thus, we need solutions that are science-based, culturally sensitive, and market-driven. The private producers will only change farming practices when there are social and economic incentives. Users of the landscape must be convinced that any proposed land management programs are not only environmentally sound but economical as well. When there is no realistic return on the inputs then the management programs will not be adopted. One way to do this is to make consumers share the burden of the cost for environmental management.

The management of prairie ecosystems should take into account the spatial and temporal structural variability of ecosystem. This means allowing processes such as grazing, burning, flooding, and water stress, so they occur cyclically over as much of the landscape as possible. This will allow the complexity of the higher-level ecosystem to be maintained. Furthermore, the landscape should be managed to maximize ecological functions, not just for economic gain, societal welfare, or threatened species or spaces. Measures of ecological function need to be developed. For example, a region could be assessed in terms of specific area-sensitive species and biotic assemblages that have dominant or keystone functions. Furthermore, the ecosystem's condition or integrity may be related to its functional diversity, which may be monitored using trends in animal guilds. To measure the effectiveness of ecosystem management for society, broad indicators should be used that include each type of capital (natural, manufactured, human, and social). This could put us on a more sustainable path as we continue to live on, and use the resources of, the northern prairie landscape.

Acknowledgments

This work was developed during many discussions over beverages with Lisa Dale-Burnett, Peter Leavitt, Michele Masley, Ken McKinney, Brian Mlazgar, Alec Paul, Dave Sauchyn, and Alan Smith. The manuscript benefited from the comments of Glenn Sutter and one anonymous reviewer.

Literature Cited

Acton, D.F. and L.J. Gregorich (eds.). 1995. *The Health of Our Soils: Towards Sustainable Agriculture in Canada.* Ottawa: Centre for Land and Biological Research.

Bakker, J.P. and F. Berendse. 1999. "Constraints in the Restoration of Ecological Diversity in Grassland and Heathland Communities," *Trends in Ecology and Evolution* 14: 63–68.

Belcher, M. 1958. "Bird Notes from a Farm Shelterbelt," *Blue Jay* 16: 101–104.

Boehm, M.M., S. Kulshreshtha, R.L. Dejardins and B. Junkins. 2000. "Carbon Sequestration and Agricultural Greenhouse Gas Emissions in Canada Based on CEEMA Analysis," *Prairie Forum* 25: 193–200.

Canada. Statistics Canada. 1996. *Census of Agriculture, 1996.* CD-ROM Release 2.1 Identifier, No. 93F0031XCB.

Christian, J.M. and S.D. Wilson. 1999. "Long-term Ecosystem Impacts of an Introduced Grass in the Northern Great Plains," *Ecology* 80: 2397–2407

Cody, M.L. 1985. "Habitat Selection in Grassland and Open-country Birds." Pp. 191–226 in M.L. Cody (ed.), *Habitat Selection in Birds.* Orlando: Academic Press.

Costanza, R., R. d'Arge, R. de Groot, S. Farber, M. Grasso, B. Hannon, K. Limburg, S. Naeem, R.V. O'Neill and J. Paurelo. 1997. "The Value of the World's Ecosystem Services and Natural Capital," *Nature* 387: 253–260.

Dahl, T.E. 1990. *Wetland Losses in the United States, 1780s to 1980s.* Washington, DC: U.S. Department of the Interior, Fish and Wildlife Service.

Davis, S.K. and D.C. Duncan. 1999. "Grassland Songbird Occurrences in Native and Crested Wheatgrass of Southern Saskatchewan," *Studies in Avian Biology* 19: 211–218.

Dixit, A.S., R.I. Hall, P.R. Leavitt, J.P. Smol and R. Quinlan. 2000. "Effects of Sequential Depositional Basins on Lake Response to Urban and Agricultural Pollution: A Paleoecological Analysis of the Qu'Appelle Valley, Saskatchewan, Canada," *Freshwat. Biol.* 43: 319–338.

Franzen, D. 2003. "Managing Saline Soils in North Dakota." North Dakota State University Extension Service Rep SF-1087(revised), Fargo, ND.

Friesen, L., M.D. Cadman and R.J. MacKay. 1999. "Nesting Success of Neotropical Migrant Songbirds in a Highly Fragmented Landscape," *Conservation Biology* 13: 338–346.

Harker, D.B., K. Bolton, L. Townley-Smith and B. Bristol. 1997. *A Prairie-wide Perspective on Nonpoint Agricultural Effects on Water Quality*. Regina, SK: Prairie Farm Rehabilitation Administration, Agricultural and Agri-Food Canada.

Herkert, J.R. 1994. "The Effects of Habitat Fragmentation on Midwestern Grassland Bird Communities," *Ecological Applications* 4: 461–471.

Houston, C.S. and J.K. Schmutz. 1999. "Changes in Bird Populations on Canadian Grasslands," *Studies in Avian Biology* 19: 86–94.

Igl, L.D. and D.H. Johnson 1997. "Changes in Breeding Bird Populations in North Dakota: 1967 to 1992–93," *Auk* 114: 74–92.

Johnson, K.H., K.A. Vogt, H.J. Clark, O.J. Schmitz and D.J. Vogt. 1996. "Biodiversity and the Productivity and Stability of Ecosystems," *Trends in Ecology and Evolution* 11: 372–377.

Leach, M.K. and T.J. Givnish. 1996. "Ecological Determinants of Species Loss in Remnant Prairies," *Science* 273: 1555–1558.

Lesica, P. and T.H. DeLuca. 1996. "Long-term Harmful Effects of Crested Wheatgrass on Great Plains Grassland Ecosystem," *Journal of Soil and Water Conservation* 51: 408–409.

Matson, P.A., W.J. Parton, A.G. Power and M.J. Swift. 1997. "Agricultural Intensification and Ecosystem Processes," *Science* 277:504–509.

Mengel, R.M. 1970. "The North American Central Plains as an Isolating Agent in Bird Speciation." Pp. 279–340 in W. Dort Jr. and J.K. Jones, Jr. (eds.), *Pleistocene and Recent Environments of the Central Great Plains*. Lawrence: University of Kansas Press.

Peltzer, D. A. 2000. "Ecology and Ecosystem Functions of Native Prairie and Tame Grasslands in the Northern Great Plains." Pp. 59–72 in T.A. Radenbaugh and P.C. Douaud (eds.), *Changing Prairie Landscapes*. Regina: Canadian Plains Research Center.

Pepper, J. 1999. "Diversity and Community Assemblage of Ground-dwelling Beetles and Spiders on Fragmented Grasslands of Southern Saskatchewan." M.Sc. thesis, University of Regina.

Persson L., J. Bengtsson, B.A. Menge and M.E. Power. 1996. "Productivity and Consumer Regulation: Concepts, Patterns, and Mechanisms." Pp. 396–434 in G.A. Polis and K.O. Winemiller (eds.), *Food Webs: Intergration of Pattern and Dynamics*. Toronto: Chapman and Hall.

Radenbaugh, T.A. 1998. "Saskatchewan's Prairie Plant Assemblages: A Hierarchical Approach," *Prairie Forum* 23: 31–47.

——. 2003. "Ecosystem Level Functional Changes in Breeding Bird Guilds in the Mixed Grassland since Agricultural Settlement." Pp. 1117–1202 in D. Rapport, W. Lasley, D. Rolston, O. Nielsen, C. Qualset, and A. Damania (eds.), *Managing for Ecosystem Health*. New York: CRC/Lewis Press.

Rapport, D.J., H.A. Reiger and T.C. Hutchinson. 1985. "Ecosystem Behavior under Stress," *American Naturalist* 125: 617–640.

Risser, P.G., E.C. Birney, H.D. Blocker, S.W. May, W.J. Parton and J.A. Wiens. 1981. *The True Prairie Ecosystem*. Stroudsburg, PA: Hutchinson Ross Publishers.

Rohrbaugh, R.W., Jr., D.L. Reinking, D.H. Wolfe, S.K. Sherrod and M.A. Jenkins. 1999. "Effects of Prescribed Burning and Grazing on Nesting and Reproductive Success of Three Grassland Passerine Species in Tallgrass Prairie," *Studies in Avian Biology* 19: 165–170.

Samson, F.B. and F.L. Knopf (eds.). 1996. *Prairie Conservation: Preserving North America's Most Endangered Ecosystem*. Washington, DC: Island Press.

Saskatchewan Conservation Data Centre. 2004. Interim List: Species at Risk Requiring Special Management Consideration. http://www.biodiversity.sk.ca

Sauer, J.R., J.E. Hines, I. Thomas, J. Fallon and G. Gough. 2000. The North American Breeding Bird Survey, Results and Analysis 1966–1999. Version 98.1. Laurel, MD: USGS Patuxent Wildlife Research Center.

Selby, C.J. and M.J. Santry. 1996. *A National Ecological Framework for Canada: Data Model, Database and Programs*. Ottawa/Hull: Agriculture and Agri-Food Canada, Research Branch, Center for Land and Biological Resources Research and Environment Canada, State of the Environment Directorate, Ecozone Analysis Branch.

Simberloff, D.S. and N. Gotelli. 1984. "Effects of Insularisation on Plant Species Richness in the Prairie-Forest Eotone," *Biological Conservation* 29: 27–46.

Smith, A. R. and T. A. Radenbaugh. 2000. "Historical and Recent Trends in the Avifauna of Saskatchewan's Prairie Ecozone," *Prairie Forum* 23:83–106.

Sutter, G.C. and R.M. Brigham. 1998. "Avifauna and Habitat Changes Resulting from Conversion of Native Prairie to Crested Wheatgrass: Patters at Songbird Community and Species Levels," *Canadian Journal of Zoology* 76: 869–875.

Thorpe, J. and B. Godwin. 1999. "Threats to Biodiversity in Saskatchewan: Plant Ecology Section." Saskatchewan Research Council, Environment Branch, Saskatoon, SK. SRC Publication No. 11158-1C99

US Fish and Wildlife Service. 1995. *North Dakota's Federally Listed Endangered, Threatened, and Candidate Species—1995.* Bismarck, ND: US Fish and Wildlife Service.

Wiens, J.A. 1973. "Pattern and Process in Grassland Bird Communities," *Ecological Monographs* 43: 237–270.

———. 1974. "Climate Instability and the 'Ecological Saturation' of Bird Communities in North American Grasslands," *Condor* 76: 385–400.

———. 1989. *The Ecology of Bird Communities: Volume 1. Foundations and Patterns.* Cambridge: Cambridge University Press.

Wright, H.A. and A.W. Bailey. 1982. *Fire Ecology: United States and Southern Canada.* New York: John Wiley and Sons.

INDEX

A

adaptive management, 3

Agriculture and Agri-Food Canada (AAFC), 89

agriculture (*see also* agro-ecosystems; farms/farmland): deregulation, 12–13, 21; energy efficiency of, from 1936 to 1991, 28–35; historical background, 11–13, 25–27; importance of, 140–141, 147; policy implications for, 36–37

agro-ecosystems (*see also* cooperatives, consumer-producer): and desertification, 139; general strategies for, 154–156; influences on vegetation and birds, 147–151; long term effects of, 153; policies for, 56–59; and species diversity, 151–152, 156

agrochemicals, 5

analogue analysis, 126

aspen parkland, 111, 112

avifauna. *see* birds

B

bio-invasion, 93–94

biodiversity (*see also* Frenchman River Biodiversity Project; species diversity): hotspots, 108, 111–112; of livestock rangeland, 91–92; of native prairie, 80

birds: conservation, 6, 21, 63, 64, 154; effect of cultivation on, 15, 16, 17–19; societal influence on, 149–150; and species diversity, 151–152

"bottom-up" processes, 69

Bourdieu, P., 42, 43

breeding grounds, 6

Brooks, A.R., 105

Brundtland Commission (World Commission on Environment and Development, 1987), 3

buffer zones, 65, 155–156

Butala, Sharon, 1

C

Canada, Government of: and desertification, 136; and grain production, 11, 13, 14, 79; national farm policy, 25

Canadian Climate Impacts Scenarios (CCIS), 122–123, 141

capital, four-phase cycle of, 3–4

carbon sequestration, 37, 80, 90–91, 95

CBEM. *see* community-based ecosystem management

centralization, 41

cereal crops. *see* grain production

church activity, 47

climate change: and desertification, 141–142; and energy efficiency, 37; and Frenchman watershed, 72; global warming, 92–93; impact in the sand hills, 117, 124–125, 126–129; model and methodology, 122–123

community-based ecosystem management (CBEM): conflicting aims of stakeholders, 70; Frenchman River Biodiversity Project, 71–76; premise and aims of, 69–70

community organizations, 48

community-supported agricultural groups (CSA), 60

confidentiality, 73, 74–75

Conservation Reserve Program (CRP), 80, 91

conservation (*see also* wildlife conservation): and drought, 139–140, 142; effectiveness of, 6; general strategies for, 155, 156; implications for, 130, 154; of migratory birds, 21; subsidies, 57; water, 76–77; and Wood River Cooperative, 62, 63, 64

constructivism, 4

consumer-producer cooperatives. *see* cooperatives, consumer-producer

continuous cropping, 14, 19, 21

cooperatives, consumer-producer: description, 59–60; goals of Wood River, 63–64; policy strategies for Wood River, 62–63; structure, 64–66; Wood River ecosystem, 60–62

crested wheatgrass, 79, 84, 88, 91, 94

crop energy, 32–33

crop insurance, 12–13

crops. *see* grain production

"Crow", the, 11

CONTRIBUTORS

KEN BELCHER is an Associate Professor in the Department of Agricultural Economics at the University of Saskatchewan. His research activities focus broadly on questions in the areas of sustainable agriculture, environmental and resource policy, climate change and ecological economic applications.

Email: belcher@duke.usask.ca

JANA BERMAN is a graduate student in the Geography Department at the University of Saskatchewan. Her Masters research focuses on the role of participatory and experiential approaches to environmental education within the context of community-based ecosystem management in southwestern Saskatchewan.

Email: jbb650@mail.usask.ca

HARRY POLO DIAZ is a Professor in the Department of Sociology and Social Studies of the University of Regina. His areas of interest involve sustainable development and rural development, with a focus on issues related to social cohesion, social capital, adaptive capacity to climate change, and rural policies.

Email: harry.diaz@uregina.ca

ANDY HAMILTON is a research scientist specializing on Homoptera-Auchenorrhyncha at the Canadian National Collection of Insects at Agriculture Canada in Ottawa. His research focuses on biodiversity. He has described 506 new taxa in 106 scientific papers.

Email: hamiltona@agr.gc.ca

ALAN D. IWAASA joined Agriculture and Agri-Food Canada in 1999. His research centres on aspects of range and pasture management such as: extending the grazing season; grazing systems for native and seeded pastures; increasing grassland productivity, animal-plant interface relationships; and forage quality assessment.

Email: iwaasa@agr.gc.ca

PAUL JEFFERSON joined Agriculture and Agri-Food Canada in 1981. He has published on many aspects of forage research including physiology, agronomy, plant breeding and quality. His current projects include: productivity and quality of native forage, forages in crop rotations, water relations, competition effects, and dryland forage agronomy.

Email: jeffersonp@agr.gc.ca

SAM KENNEDY recently graduated from the University of Regina with a Master of Science degree in Geography. His thesis examined landscape sensitivity and climate change in southern Saskatchewan. Currently he is employed with Alberta Sustainable Development in Edson, Alberta, where he is analyzing and managing GIS data.

Email: samuel.kennedy@gov.ab.ca

JEAN LAURIAULT is an Environmental Specialist with the Canadian Museum of Nature's Centre for Biodiversity. His interests include training in biodiversity, the development of national strategies for the conservation of biodiversity, environmental education, botany, and Monarch butterfly conservation.

Email: jlauriault@mus-nature.ca

DIANE MARTZ is the Director of the Centre for Rural Studies and Enrichment at St. Peter's College in Muenster and a lecturer in the Department of Geography at the University of Saskatchewan. Her work focuses on the sustainability of rural communities, the adaptation of farm families to agricultural restructuring, and domestic violence in rural communities.

Email: martzd@stpeters.sk.ca

GRANT MCLEOD has studied the development of germplasm collections of native plant species through Agriculture and Agri-Food Canada since 1995. He has released several ecovarsÔ from that research and currently leads a major project aimed at developing native ecological varieties for reclamation, re-vegetation and agricultural production.

Email: mcleodg@agr.gc.ca

MARK NELSON is a Master's candidate in the Department of Sociology and Social Studies at the University of Regina. He recently participated in a research project on the social cohesion of rural communities, working on the development, implementation, and analysis of two telephone surveys of rural community residents.

Email: mark.nelson@uregina.ca

TODD A. RADENBAUGH is an interdisciplinary earth and environmental scientist who enjoys investigating ecosystem level processes in both fossil and modern landscapes. He has worked in Bermuda, Canada, Germany, Jamaica, Kosovo, Sweden, Turkey, and United States is continuing his work in Virginia.

Email: todd.radenbaugh@uregina.ca

GREG RIEMER is a Senior Policy Advisor to Saskatchewan Environment and a Director of several environmental non-governmental organizations. His work focuses on environmental, land-use and agriculture issues including, the NAWMP, the Saskatchewan Prairie Conservation Action Plan, Conservation Lands Policy, and the Great Sand Hills.

Email: griemer@serm.gov.sk.ca

DAVE SAUCHYN is the PARC/Manitoba Hydro Research Professor at the Prairie Adaptation Research Collaborative (PARC) and a Professor of Geography at the University of Regina. His academic interests are the climate of past millennium and the response of natural and social systems to climate change and variability.

Email: sauchyn@uregina.ca

MICHAEL SCHELLENBERG has studied forage production since joining Agriculture and AgriFood Canada in 1984. His current research examines the impacts of biodiversity on society and the agricultural industry, as well as the impacts of agriculture on biodiversity, especially native grasses, legumes, and shrubs, climate change, and invasive plants.

Email: Schellenberg@agr.gc.ca

JOE SCHMUTZ is the Conservation Planner for the Important Bird Areas program of Nature Canada and Nature Saskatchewan. He is also a Fellow in the Centre for Studies in Agriculture, Law and the Environment at the University of Saskatchewan. He has studied prairie raptors and now addresses conservation at the landscape and ecosystem level.

Email: joe.schmutz@usask.ca

ROBERT SISSONS has an M.Sc. in Wildlife Ecology from the University of Alberta based on studies of burrowing owl habitat use in southern Alberta. He is currently a Conservation Biologist for Grasslands National Park in Val Marie, Saskatchewan, where he leads the development and implementation of the ecological monitoring program.

Email: robert.sissons@pc.gc.ca.

JENNIFER STROICH is Rural Development Analyst with Agriculture and Agri-Food Canada, PFRA and a Master of Science (Geography) candidate at the University of Regina. Her research focuses on agricultural drought issues from a policy perspective, dendro-chronology as a measure of drought variability, and drought indicator application across Canada.

Email: stroichj@agr.gc.ca

GLENN SUTTER is Curator of Ornithology and Human Ecology at the Royal Saskatchewan Museum and an adjunct professor of biology at the University of Regina. His research focuses on prairie conservation issues and the resilience of complex systems, with an emphasis on grassland songbirds, ecosystem health, and sustainability education.

Email: gsutter@royalsaskmuseum.ca

BOB STIRLING is a Professor of Sociology & Social Studies, and Political Science, at the University of Regina.

Email: Bob.Stirling@uregina.ca

JEFFREY THORPE is a Senior Research Scientist at the Saskatchewan Research Council. He has an M.Sc. in plant ecology from the University of Saskatchewan, and a Ph.D. in forest ecology from Yale University. His current work focuses on the application of ecological knowledge to forestry, range management, conservation, and land-use planning.

Email: thorpe@src.sk.ca

STEPHEN WOLFE is a geomorphologist with the Geological Survey of Canada and an adjunct professor at several universities. He has studied links between past climate and dune activity on the prairies and eolian deposits in the Yukon and British Columbia. He is currently leader of a major project called Paleoenvironmental Records of Climate Change.

Email: SWolfe@NRCan.gc.ca